Surviving the Wild Wild Web

The information security arena is often vague and confusing for internet users, both young and old. New traps are being devised daily, and falling into them can take legal, ethical, financial, physical, and mental tolls on individuals. With increasing cases of fake news, identity theft, piracy, spying, and scams surfacing, this book explains the risks of the internet and how they can be mitigated from a personal and professional perspective.

Surviving the Wild Wild Web: A User's Playbook to Navigating the Internet's Trickiest Terrains is a readable guide addressing the malicious behaviors within internet cultures. Written in simple and jargon-free language, the book describes ten pillars of information security risks faced by all internet users. Each pillar will be detailed as a story, starting with the roots of the problem and branching out into tangential related issues and topics. Each chapter ends by detailing ways a user can avoid falling victim to cyber threats. It uses a combination of news articles, topical current events, and previously published academic research to underpin the ideas and navigates how users interact with the World Wide Web. The book aims to create a generation of internet-literate readers who can spot the pitfalls of the internet in their personal and professional lives to surf the web safely.

This guide will appeal to any individual interested in internet safety, with a potential readership extending to students and professionals in the fields of computer science, information systems, cybersecurity, business, management, human resources, psychology, medicine, education, law, and policy.

Surviving the
Wild Wild Web
A User's Playbook to Navigating
the Internet's Trickiest Terrains

Marton Gergely
Ian Grey
Heba Saleous

CRC Press
Taylor & Francis Group
Boca Raton London New York

CRC Press is an imprint of the
Taylor & Francis Group, an **informa** business

Front cover image: F01 PHOTO/ShutterStock

First edition published 2025
by CRC Press
2385 NW Executive Center Drive, Suite 320, Boca Raton FL 33431

and by CRC Press
4 Park Square, Milton Park, Abingdon, Oxon, OX14 4RN

CRC Press is an imprint of Taylor & Francis Group, LLC

© 2025 Marton Gergely, Ian Grey and Heba Saleous

Library of Congress Cataloging-in-Publication Data
Names: Gergely, Marton (Instructor), author. | Grey, Ian (Associate professor), author. | Saleous, Heba, author.
Title: Surviving the wild wild web : a user's playbook to navigating the internet's trickiest terrains / Marton Gergely, Ian Grey and Heba Saleous.
Description: First edition. | Boca Raton, FL : CRC Press, 2025. | Includes bibliographical references and index. | Identifiers: LCCN 2024033651 (print) | LCCN 2024033652 (ebook) | ISBN 9781032679372 (hardback) | ISBN 9781032645124 (paperback) | ISBN 9781032679389 (ebook)
Subjects: LCSH: Internet--Social aspects--United States. | World Wide Web--Social aspects--United States.
Classification: LCC HM851 .G472 2025 (print) | LCC HM851 (ebook) | DDC 302.23/10973--dc23/eng/20241106
LC record available at https://lccn.loc.gov/2024033651
LC ebook record available at https://lccn.loc.gov/2024033652

ISBN: 978-1-032-67937-2 (hbk)
ISBN: 978-1-032-64512-4 (pbk)
ISBN: 978-1-032-67938-9 (ebk)

DOI: 10.1201/9781032679389

Typeset in Times
by SPi Technologies India Pvt Ltd (Straive)

Contents

About the Authors

Marton Gergely teaches in the Department of Information Systems and Security at the United Arab Emirates University. He was born in New Delhi and grew up in Budapest. He holds a PhD in Business Administration, with an emphasis in Information Technology, from The University of Texas at San Antonio. His current research interests include digital piracy, social and cognitive psychology in technology use, cyber law and ethics, social desirability bias, as well as emerging technologies.

Dr. Ian Grey was born and raised in the wilds of Co. Waterford, Ireland. He holds a B.A. and a Ph.D. in psychology from the National University of Ireland (Cork) and a D. Clin. Psych from Trinity College Dublin. He holds additional postgraduate qualifications in Applied Behavior Analysis and in Forensic Psychology. He is currently an Associate Professor of Clinical Psychology in the Department of Cognitive Science, at the United Arab Emirates University where he teaches courses on clinical psychology practice. His research interests include cognitive vulnerabilities, neuro-developmental disorders and determinants of mental health.

Heba Saleous is a Palestinian-American currently working toward earning her PhD in Information Security at the United Arab Emirates University under the supervision of Dr. Marton Gergely. She holds an MSc in Information Security from the same institution. Her current research interests include digital forensics, video game forensics and security, cybercrime and terrorism, cybercriminal behavior, and malware.

Introduction

You almost certainly have never heard of a man named Thomas Midgley, and were he still alive today, he would probably wish that his name remained confined to the shadows of history. Midgely was an American inventor, who in his lifetime, won several major scientific prizes, had over 100 patents for his wide-ranging inventions, and was elected to the United States National Academy of Sciences, which is about as big an honor as a scientist can get. Yet, he also has the unfortunate moniker of being credited as being the one man whose actions have resulted in more environmental disaster than anyone before him or since. You can be sure that Greta Thunberg will not be posting pictures of him any time soon.

In the 1950s, Thomas was tasked with how to make refrigerators safer and less likely to explode in the kitchen, which they apparently had a habit of doing at the time. Thomas correctly identified that the chemicals being used were the source of the problem and in response he invented what are known today as chlorofluorocarbons, or CFCs, which subsequently went on to be used in billions of aerosol cans and even asthma inhalers. He also invented leaded petrol to solve the problem of engine knocking, which was a feature of cars of the day and made millions of dollars for the company he worked for. There was only one issue that no one at the time seemed to reflect on: the law of unintended consequences. Thomas was by no means a bad guy, just a scientist with a knack for invention, but CFCs had the consequence of depleting the ozone layer high up in the sky, and the use of leaded petrol released massive quantities of lead into the atmosphere all over the world. Both inventions have contributed to greenhouse gases and the existential threat posed by climate change today. Unfortunately, due to the internet, you might think

DOI: 10.1201/9781032679389-1

that's fake news. Former President Donald Trump certainly thinks so, as you will discover in the following pages.

The cautionary tale from the story of Thomas Midgley is that we humans are very, very good at inventing things, but not that good at predicting how things might affect us down the line. Even if we do think of them, we also have a human tendency to grossly underestimate the potential negative consequences and get hijacked by the upside. A bit like dating a supermodel with narcissistic personality disorder. Sure, they may lack empathy, require constant attention and admiration, and just happen to be exceptionally arrogant... but, damn they're gorgeous. It's all gonna be totally fine. Until it's not. Just look at the recent invention of artificial intelligence. Wonderful things are being predicted about its use in the fields of business, medicine, and logistics, along with the promise of enhanced decision-making. Mention student essays or the plot of the Terminator movies and suddenly, you're a total killjoy. The long history of human stupidity should make us all a little more skeptical and wary. Except, we won't, and even a cursory awareness of the history and contemporary characteristics and consequences of the internet would make you aware of that.

The internet was invented in 1983 and 10 years later in 1993, the World Wide Web was released by its inventor, Tim Berners-Lee, into the public domain. It completely revolutionized the internet and allowed ordinary people who were not computer scientists to create things called websites and fill them with text, graphics, audio, and hyperlinks. Essentially, there were three broad predictions made back in 1993 about what the future of the internet would look like. The first was that there would be no future for the internet. Robert Metcalfe, the inventor of Ethernet, predicted that the internet would "go spectacularly supernova and [...] catastrophically collapse" within two years [1]. Being a betting man, he promised to eat his words if his prediction did not come to pass. Two years after his prophecy, while giving a keynote speech at an international conference on the World Wide

Web, he publicly took a printed copy of his internet column that predicted the collapse, put it in a blender with some water, and then consumed the "eat your own words" smoothie. You have to admire a man true to his word.

The second of these premonitions was that because the internet allowed lightning-fast information to your fingertips, future generations would become the smartest in the entire history of humanity, skipping home from school to plow into Encyclopaedia Brittanica for the evening with the rest of the family. The idealism extended as far as people even believing it would lead to a new form of social utopia by virtue of connected and expanded communities. The American poet John Perry Barlow went so far as to wax lyrical about the social possibilities of the internet, when he wrote back in 1996 that "most human exchange will be virtual rather than physical, consisting not of stuff but the stuff of which dreams are made" [2]. Except, John would have known that some dreams are, in fact, nightmares.

The last prediction was perfectly captured by a lady called Elizabeth Vargas, who was an anchor on the American channel NBC back in 1993. She said to her co-host "Wouldn't it bother you that there are all these people you don't really know? Everyone is signing on, having these conversations, whining together or griping together…" [3]. Now, take a moment and ask yourself which of these predictions seems to best reflect the current incarnation of the internet that we are living with. What percentage of kids are in fact coming home to read the encyclopedia, and even more importantly what are the rest of them doing? Are our virtual social worlds now reflective of beautifully benign dreams? How much whining and griping is currently going on? This book explores those consequences, and indeed had they been limited to simply enhanced griping, it might have turned out better for many of us.

The internet is now omnipresent and woven into the very fabric of our lives. In some ways, it has been brilliant. It has given us the ability to connect with friends and families across vast distances, the ability to easily perform the boring but necessary life tasks of paying bills and also the means to book our next adventure holiday without getting off the couch (which is a bit ironic). It has also enabled you, if you are so inclined, to spend months on end in the confines of your own home ordering what you need online, as a scary number of Japanese people are currently opting to do. That's all great but the internet has also become a hot mess. It's a digital world where both the vulnerabilities and darker sides of humanity get exploited and played out. It's become a minefield and something that should be approached with the level of trepidation one would associate with approaching an open cesspool. It has its uses, but also its dangers.

The sad reality is that many of us don't go on the internet to peruse at our leisure the entire collective wisdom and knowledge of humanity. Instead of making us smarter, we have largely outsourced our ability to think to internet influencers and artificial intelligence. We have come to rely on algorithms to find love and connection. We spend our leisure time sitting next to our nearest and dearest in our own world watching TikTok videos of cats being cats, people slipping on ice, and catching up on celebrity gossip. We worry about who our partners are meeting online, while they are in the bathroom. We have gotten bogged down in echo chambers of our own beliefs, fueled by fake news and online communities, which merely reflect back to us our own thoughts. We get exposed to scams, spying, data and identity theft. We get manipulated and often we are not even aware of it. The children who have grown up in a world surrounded by the internet since birth have not become the happiest generation, despite our technological advances and associated promises. Just ask the parents of the current generation. They will tell you that they are exasperated with, confused for, and worried about their kids. They are right to worry, given how the way social media platforms are intentionally designed, and what content is promoted within. The internet has not proven to be the panacea to our weaknesses and limitations. In some unfortunate ways, it has hijacked and amplified them, and often for the benefit of others.

We have specifically decided to write this book NOT as an academic text. While the three of us are in fact academics, to be honest, we find that approach boring, and reading rigor mortis-inducing text will cause you to stiffen up. So, we unanimously chose to avoid that (I'm sure at our publisher's dismay). Nor is this book intended to be a luddite-inspired rant about the internet, though it might come across that way (and there most certainly is plenty of ranting). We like to think of the internet as that quirky friend who is both amazing and infuriating at the same time. It's a place where serious news stories can sit side by side with absolute nonsense, like Barbie makeup tutorials, the Hawk Tuah girl, and the Ozempic craze. A place where your perfectly ordinary search for a banana bread recipe can somehow, magically, lead you to a conspiracy theory about aliens currently inhabiting the White House. It's a place where we're all connected, yet strangely at the same time somehow more isolated than ever, with our noses pressed up against screens, scrolling through an infinite feed of information and dribble. Like any good therapist will tell you about life, it's complicated, and so is understanding how the internet has affected our lives.

We are not going to talk about the benign aspects of the internet here. There is no point in talking about the benefits of having a footpath when you are about to stumble into an open manhole or find yourself at the bottom of one needing to get out. Instead, we have written this book as a guide to surviving the internet, by describing the dodgy stuff that takes place, and as a means of becoming more aware of the trip-wires for potential harm. We do not expect an endorsement of this book from Donald Trump, Mark Zuckerberg, or the Chinese owners of TikTok in particular. But just like the internet, and especially like your beloved social media platforms, we want to keep you engaged to find out exactly why. Buckle up! You can decide to turn off the router later. As a last resort, we can also recommend a good Zen monastery in the Catskill Mountains.

REFERENCES

[1] S. Clarke, "25 years on, here are the worst ever predictions about the Internet," New Statesman, Aug. 23, 2016. [Online]. Available: https://www.newstatesman.com/science-tech/2016/08/25-years-here-are-worst-ever-predictions-about-internet. [Accessed: Jun. 27, 2024].

[2] J. P. Barlow, "John Perry Barlow," Elon University, 1990. [Online]. Available: https://www.elon.edu/u/imagining/time-capsule/early-90s/john-perry-barlow/. [Accessed: Jun. 27, 2024].

[3] S. L. Clarke, "How the promise of the Internet was a lie," Medium, Oct. 28, 2021. [Online]. Available: https://medium.com/@seanlandonclarke/how-the-promise-of-the-internet-was-a-lie-2c33736cbb6e. [Accessed: Jun. 27, 2024].

1 Pirates of the Ethernet – Adventures in Illegal Downloading

"Bee boop, beep boo, beep boop. Eeeee urrrrr, deedle dee-dle, screeeellllllll, ee ee ee, burrrrrr blur blur."

That is either music to your ears, or nails on a chalkboard. Nothing in between. Nonetheless, that is the sound of a lot of our youth. Some of you (my dear wife included, who most likely will never even open this book) have no idea what I'm talking about. You are the children of the nineties and beyond. But for those of you fortunate enough to be born before then, you will know this sound well. The gateway to your freedom. The gateway to Valhalla. The gateway to the World (Wide Web).

We used to sit at our 1990s Compaq computers and wait for this sound. But only when everyone else was fast asleep, and the phone lines were free. Booting up the free AOL trial software that we would take by the handfuls at the local Blockbuster video (conveniently given away on a pretty little CD with the ubiquitous little yellow AOL man running on the cover). Then we would fire up the godsend that was Napster. Searching for that new Blink-182 song we just heard on the radio, or even better, saw on MTV. We would wait patiently as the blue progress bar started eclipsing the gray horizontal rectangle, knowing we

DOI: 10.1201/9781032679389-2

were only a good 15 minutes away from pure aural bliss. Once you got enough of these downloaded, you could even burn your own playlist to a CD, and play it for your friends on the school bus. But you have to be careful and hold it still, as those Discmans aren't skip-proof yet.

If the above scenario reminded you of the good ol' days, you were probably also of the cassette tape era. Waiting patiently by a radio for a good song to come on, so you could quickly press the record button on your AIWA boombox, hoping the incessant chatter of the DJ wouldn't interfere with your perfectly coordinated mix tape. Or perhaps you are more of a visual being and were obsessed with recording Beverly Hills 90210 onto a Sony Dynamicron E-180 blank tape. Nonetheless, this was all a different time. In all of these situations, we weren't as connected. Pre-mobile phones, pre-social media, we didn't put ourselves out there as much. The world has since changed, as have the ways we indulge in media. And with every advancement of technology, every new innovation, around the corner, lurking in the dark, lie the downsides. But before we go full-blown Pessimistic Patty, let's unfold how and where these copyright-infringing acts, otherwise known as piracy, stem from.

Whoosh! You're back in time. Easter 1770 to be exact. A young boy and his father are visiting the Sistine Chapel, and hear a song being sung by the choir. The song is Gregorio Allegri's "Miserere Mei, Deus", composed in 1638, and an immensely close-guarded 150-year-old secret by the Vatican. It was only ever to be sung during Holy Week, and only within the hallowed halls of the papal chapel. The music was never to leave the chapel and the sharing of it strictly forbidden, under threats of excommunication [1]. In fact, the piece's sheet-music was only authorized to be owned by three people: Leopold I (The Holy Roman Emperor), John V (The reigning King of Portugal), and Giovanni Battista Martini (an Italian Friar and composer) [2]. Anyways, this young boy, just 14 years old, is enamored by the song, and can't get it out of his head. So much so, that he transcribes the entire piece from memory, and this transcription ends up being published in London the following year. The young boy? Mozart.

Okay… whether this is true or not, no one can be one hundred percent certain, but you have to admit, the story itself is pretty cool. Some argue that the ability of a child (even of Mozart's caliber) to transcribe a piece of music from memory after hearing it only once (or maybe twice), is highly implausible. Others say that the piece he wrote was not, in fact, the same piece as sung by the Sistine Chapel choir, but just a very similar composition heavily inspired by it (I guess the term "cover" or "sample" wasn't invented yet). Either way, we have a very interesting beginning to our story about illegal downloading, if I do say so myself.

Fast forward a few centuries to the 1980s and we suddenly have Cher's perm, a-ha's hit single "Take on Me", the DeLorean DMC-12 (and Marty McFly), K-Swiss neon windbreakers, and two-tone steel and gold Rolex Datejusts. And we also got the era of bootlegs. And no, we're not talking about the shaft of the boot where cowboys in the 1800s would stash their illicit weapons and booze to keep them out of sight (although that is where the term originated). We are, of course, talking about illegally copied music, movies, and software, in other words, the "big three" of piracy. Let's break each one down individually.

MUSIC

Since we already mentioned our good friend Wolfgang (Amadeus, Amadeus), it's only appropriate that we continue with music first. So, what have been the big technological marvels that allowed music to be so freely and blatantly copied? Well, first it had to be the advent of the transistor radio in 1948. Without this neat little gadget, people would not have been able to wirelessly listen to their favorite songs at home, at work, and on the move. Prior, music was largely enjoyed on vinyl, and as we all know, vinyl is not exactly portable. So, along came radio to save the day. But, this alone did no harm! In fact, it only helped to popularize music more than ever before. Radio stations paid royalties to musicians, and everyone was happy. Until the people over at Philips in the Netherlands came along in the 60s and really screwed everything up... but we're getting ahead of ourselves a bit. By inventing the Compact Cassette, otherwise known as the cassette tape, people were now able to buy pre-recorded "records" that were robust (relatively, you may need a pencil... brownie points for those that get this reference!), and most importantly, easy to transport anywhere you go. Essentially, these tiny spools of magnetic tape worked very much like that larger-scale reel-to-reel versions that radio stations used to play their songs. Of course, those were huge and bulky and needed specialists to operate. So along comes the wee-little cassette tape, a revolution in its own right. Not only can anyone play songs without the need for an advanced degree in audio engineering, but they can also record things!! "What?!?!", I hear you scream. I know. Mind. Blown.

Originally, the ability to record on tapes was merely through a microphone. This allowed people to record live music and entire concert sets, as well as make the ubiquitous less-than-ideal copy of a song playing on the radio, or perhaps copy of another tape. I told you those guys at Philips were the real evil geniuses behind music piracy (of course, I kid). Now compound this technology with the rise of boomboxes (where you have a radio AND a cassette player in one), and the rise of the icon that is the Sony Walkman, and all of a sudden we have a perfect storm for piracy. Essentially, not only was copyright infringement practically encouraged, it was built in! Users could simply record songs from the radio creating their very own mixtape (which you can then give to your boo), or better yet, with a second tape deck (which some boomboxes already came with) you could record tape-to-tape, and duplicate your friend Nigel's entire collection of albums with almost zero effort. Perfect, now you could take your bootleg copy of N.W.A's "Straight Outta Compton" with you wherever you go, with your handy-dandy Walkman.

At one point, in the early 1980s, this illegal copying got so bad that the British radio stations started playing a "Home Taping Is Killing Music" slogan to attempt to fight copyright infringement [3]. One cassette version of the Dead Kennedys' "In God We Trust, Inc." album even featured a blank side, printed with the message "Home taping is killing record industry profits! We left this side blank so you can help" [4].

As the years passed, things began to "digitize", cassette tapes gave way to the CD. No more analog tape, therefore no more easy recording. Or so we thought. Sure, with a CD you don't have the in-built ability to make mixtapes from a boombox, but with every hurdle in copyright protection, a new door opens. Enter the world of the so-called "CD burner". Through the use of this little gadget, we were essentially able to duplicate audio CDs. First, we "ripped" them onto our hard drives as this new-fangled file type called an MP3, and then we copied them to our new best friend, the Maxell 700MB CD-R. Audio quality also got a massive boost over tape tech! Win, win! Combine this with a label printer and a blank jewel case and BOOM! You're now selling bootleg copies of Daft Punk's Homework for $3 a pop to the 10th graders at lunchtime. But, of course, one problem still loomed. Sharing these individual ripped MP3 files was troublesome. We didn't have terabytes in our pockets in the form of external drives and USBs, not to mention the super long read/write times (copy-pasting back then took ages). What are we to do? "Ding-dong"... Napster has entered the chat.

We called it Peer-to-Peer (P2P) file sharing. The year was 1999. American Pie, Fight Club, The Matrix, Blair Witch Project, 10 Things I Hate About You. Ok, did I set the scene? Emily has just bought a CD of OutKast's new album Stankonia (the one with Sorry Ms. Jackson and So Fresh and So Clean, Clean) at her local Best Buy over on Westheimer. Her friend Jenny really wants it too, so Emily, being the nice girl that she is, rips the songs to her PC (with that blazing fast Windows 98 OS), and grabs a blank CD from the spindle of Imation CD-Rs. Done and dusted. Now Jenny has a copy of Stankonia too! But what happens to those MP3s left behind on Emily's computer? Does she delete them? NO! Of course not! Maybe they will come in handy later. Well, little did Emily know that when she had installed Napster a few weeks earlier, that 'Music' folder, where she rips her CDs to, became the "Shared" folder for her Napster software. So essentially, anytime Emily goes to download a new song, she is simultaneously uploading the Stankonia album, song by song, to potentially millions of users worldwide.

Well, that didn't last long... Napster launched in June of 1999, but by October 2000 they already had heaps of legal troubles [5, 6]. Eventually, nearly two years after its inception, Napster was shut down by court order, at the same time, defining and guiding the future rules of online downloading along the way. The problem was simple. While Napster could be used for legal things, for instance sharing that PowerPoint presentation for your Econ 101 class with that one group member Bryan who never contributes anything, but still always manages to get an "A", that's not what it ended up being used for. Napster had the ability to make sure illegal copyright-infringing activities were not happening on their watch. In fact, quite the opposite, they kept a central directory of all the files that people were sharing. And that's a big no-no. This caught them red-handed, and they couldn't plead ignorance. Ergo, Napster was found liable for copyright infringement and thus forced to shut their doors (at least in their original guise) [7, 8].

Afterwards came the deluge of second-gen P2P programs, all doing the same thing. But legally speaking, doing something TOTALLY different... because they didn't retain a centralized repository of files that passes through their systems [7]. Same same, but different (at least in the eyes of the law). LimeWire, Gnutella, Morpheus, Kazaa, and most importantly, BitTorrent, all learned from Napster's mistakes, and quickly exploited the court ruling's glaring loophole, and enabled a DHT (distributed hash table) architecture, stored on the clients' end, to record the location and availability of files and to coordinate users [9]. Essentially, as BitTorrent evolved into an interdependent network of countless users, regardless of which user is shut down by legal action, the entity itself continues to live on. Now add to this the increase in internet speed globally, and now you're able to download entire albums in seconds, as opposed to just songs in minutes. As such, the scale of the whole illegal downloading music problem blew up uncontrollably [3].

Artists were not being paid for their music, record labels were losing money hand over fist, something had to be done. So, music and tech companies got together to create a new age of music. Essentially, the rise of digital music players (most importantly a little white one that had a scrolly wheel on the front), shifted people away from CDs. At this point, there were two ways to get music. One was the legal way… to go to a physical store and purchase a CD. The second was the choice everyone made… either getting your friend Emily to burn you a CD, or just downloading it yourself online. So, the big revelation that record companies had to ask themselves was: why were CDs even necessary if all anyone was going to do was rip their songs from a physical disc to their computer anyway? Why not completely eliminate the middleman? And that is how the iTunes store was born. Some of you may remember such gems as WinAmp, but these only allowed you to play your MP3s and WAV files that you downloaded or ripped from other sources. Hence the genius of SELLING digital downloads natively within the iTunes ecosystem. Essentially, what Apple did for iPhone software in 2008 (we'll cover that later when we talk about software), they did first for music in 2003. Now, not only could you bypass the pain in the ass that was waiting to rip a CD (although in all fairness, with advances in computing tech they did get faster and faster), you didn't even have to leave your house! Furthermore, you didn't even have to buy the whole album, as with those damn physical CDs. We already know that 90% of the songs on any given album are just filler anyway. Now you could buy just the individual songs you wanted! For a fraction of the price of the whole album! Win, win! AGAIN!

But one major problem loomed. What's to stop someone from purchasing a song on iTunes, and immediately sharing it online through BitTorrent or any of the countless apps we talked about earlier? Did the record and tech giant make this big of a mistake? Of course not! Welcome to the realm of DRM. Digital Rights Management enabled music creators to protect their songs from being shared. Furthermore, when all the other tech companies jumped on the iTunes store bandwagon, everyone had their own file types too! Sony, Windows, Amazon, etc… and none of them were compatible with each other. Naturally, this made everyone very happy and peaceful. You may have legitimately paid for a song on iTunes, but couldn't upload it to your Microsoft Zune. Big sigh. Eternal bliss. No, but really, not only was the content different file types but it was also encrypted. So, send it to anyone you like, upload it anywhere. But without the licenses and digital signatures (which were insanely more complicated to transfer), the music files became useless.

Over time, methods to crack these DRMs emerged, and tech companies slowly started to ease up and morph all of the file types into a few ubiquitous ones. Further, with the advent of streaming platforms like Spotify and Deezer, the idea of even "owning" (even free) music was becoming less and less interesting. Between 2017 and 2021, there was a 65% decrease in music-related piracy globally, and most of this decrease is attributed to the rise of streaming [10]. We also began to see many artists simply uploading their own songs to YouTube, their websites, and other similar platforms, free of charge. Artists letting go of this idea of ownership, in my opinion at least, seems to be the path moving forward. Concentrating on license agreements with streaming providers, radio stations or with films and TV shows (have you noticed how insanely bangin' most Netflix show's soundtracks are these days? I'm sure they pay big money for the music rights), as well as on ad revenues from websites or YouTube, while simultaneously relying on concert ticket and merch sales are where it's at. Sure, musicians these days can still try selling songs and albums on the likes of iTunes, and even sell some CDs and vinyls for the really old-school folks out there, but in the long run, this is not going to be the bread and butter of the musician's paycheck moving forward, and the quicker that is understood and accepted, the better off we will all be.

Now lastly, you may hear some chatter about how post-2021 the music piracy numbers have started to increase again with the major culprit being the so-called "stream-ripping" [10]. Simply put, there are hundreds of sites and apps that you can find with a simple Google search that allows you to essentially download anything from the likes of Spotify or even YouTube into a nice handy file, and make it publicly available to share with eight billion of your closest friends [11]. Now, with every step toward eliminating music piracy (or any other type of piracy in general) we will always face some type of setback, as tech-savvy individuals find ways to counteract the measures. Nonetheless, one thing that we should always remember is, we have to be very weary when we hear blanket statements like:

> In just the single month of July 2019, Ed Sheeran's album Divide had over 612K downloads, Kanye's The Life of Pablo (2016) had 280k and Lady Gaga's debut The Fame Monster (2009) had over 202k.
>
> Three albums, picked at random, being illegally downloaded over a million times a month.
>
> Based on a typical iTunes or Amazon download retail price, this represents approximately $10m of lost revenue for the music business and its retailers.
>
> [12]

Just because over a million people downloaded these albums, DOES NOT constitute $10 million in lost revenue. This type of statement hinges on one giant fallacy: that EVERYONE that illegally downloads a song or album would have purchased it with their own real-life money had the opportunity to not illegally download it not been present. This is absolutely not true, but music companies would love for you to believe it. I will end this section with one final example of this. India is currently ranked number one at the moment in music piracy [10]. The annual per capita income of India at the time of writing this is right about $2,000 (and that's after it doubled since 2015) [13]. The average price of an album on iTunes is $9.99 [14]. Do you think this computes? Sorry to say Republic Records, but Manoj the chaiwala from Uttar Pradesh isn't about to drop his monthly income on Taylor Swift's newest album, The Tortured Poet's Department. Just chalk it up as free advertising.

FILM AND TV

Ok, so we spent the majority of our time here talking about music. But that's just because it was the earliest form of "piracy" with that Mozart example, and also what reminds me most of my youth. But, it is also because the tools, technologies, and trends in film and TV piracy basically mirror what happened in music. Instead of cassettes, we had Betamax and VHS. Essentially just two sides to the same coin. The giants at Sony created Betamax in 1975 and a year later JVC created VHS (by the way, did you know that JVC stands for Victor Company of Japan? And VHS stands for Video Home System. AND... VCR, which we used to record to the VHS tapes stands for Videocasette Recorder. Fun fact.). I'm sure no one copied anyone's homework, and JVC worked independently and without even glancing at a Betamax and vice versa. Actually, they both worked together on the predecessor which was called U-Matic [15], but that's neither here nor there. Ok, back to our story... we all know VHS ended up winning the battle for the home video format, and for a multitude of reasons. It was generally cheaper, more easily accessible, more compatible with different devices, etc. But one big blow to Sony's Betamax could have been its legal woes.

In 1983, Sony was sued by Universal City Studios. A case that lasted a year in courts can't be good for business, right? Just like with the Napster case, it was not about what Sony did or didn't do, as neither Sony nor Napster physically did any copyright infringing on their own. It's about what they enabled users to do. Universal City Studios argued that Sony made it easy for their Betamax users to make copies of anything that aired on TV. Now this claim was later shot down by the US Supreme Court, who ruled that it was entirely fair use for an individual to make such copies, so long as it was for personal use [16]. So in a nutshell: making a copy of an episode of Knight Rider to watch later in the comfort of your living room after a hard day's work while enjoying a Swanson's Salisbury Steak TV dinner, totally okay. But making one hundred copies of an episode of Dallas to sell on Canal Street between Church and East Broadway? Not okay. For God's sake, in India they were even running paid VHS libraries publicly, and renting out these bootlegged movie copies [17]!

Well, slowly yet surely, VHS yielded to the DVD. We're not even going to attempt to talk about the atrocity that was LaserDisc, but you're free to look that one up on your own time. Just like with CDs, we now had digital files, which meant easier copying. So here came the DRM train again, with all types of new inventions. But once again, easily stripped with the right tools. The one thing that remained in favor of the DVD for quite some time was the file sizes versus the bandwidth and Internet speed available. While songs remained fairly small, and therefore even with slow download speeds could be managed, films and shows were much, much larger. So, it was not until the rise of high-speed broadband that illegal downloading reared its ugly head. Until then, things remained pretty simple. Rip a DVD, burn a copy. Let me just say this—I may or may not remember watching the first Harry Potter movie (on a DVD that I purchased for $5 in my high school music class, no less) with a big banner across the bottom that said "Property of Warner Bros. Pictures"…

After the DVD came the HD DVD vs. Blu-Ray battle (nice to see Sony taking the W after the L that was the Betamax). These once again increased file sizes, but by this time (the mid noughties) Internet speeds were more than enough, and online downloads through P2P networks rose to an all-time high. But, just like with music, the invention of streaming platforms (here's to you Netflix, because without you, who knows where we would be) caused users to begin to forgo the idea of "ownership" again (just like with music). Up popped Hulu, Amazon, and all the "+"s. It seemed at one point every TV channel worth its salt added a "+" after their name and began charging $10 a month. This, of course, mirroring the music industry, then had its fair share of stream ripping or illegal streaming. Nowadays, for about $20, you can get a little black box out of China that you can just plug into your TV via HDMI and get every single TV channel in the world! Doesn't work very well, but it's still possible!

Lastly, we need to talk about our friend and foe, "HDCam", or sometimes just "CAM" (if you know, you know). In the dark of a movie theater, some people, will... whip out a giant tripod and a little camcorder, and record an entire film. They will then go home, and upload it to any number of sites as torrents. They're shaky, the audio's shit. But still people download them. This is a strange and unique aspect of piracy that doesn't really have an equivalent in music piracy (perhaps recording a concert, but no one would do that as audio, it would most likely be video, and then were right back here to video piracy). Nonetheless, it is still a prevalent hurdle that the film studios need to deal with.

Now, unlike with music piracy, I'm not here to make any sweeping declarations on how to fix the issue of film and TV piracy. This one is a bit tougher. Once again, in 2024, we're seeing a rise in piracy levels globally. And again, India is in the lead (but with nearly one and a half billion people, and to think that 15 years ago there were hardly five million broadband Internet connections in the country, it's not a stretch to see why) [18]. But at the same time, we should look at this globally, there is roughly a 1% increase in global population per year. At the same time, there is roughly a 6–8% increase in internet users per year, and to boot, India only currently has an internet adoption rate of about 50% [19]. And that is just India... globally internet usage is hovering right above 64%. Compounding this with the fact that movies and TV shows are generally more expensive than music, this all ends up meaning that there will be much more piracy to come in this sector, and we're all in for a very rough ride (Why we? Keep reading to find out). Obviously, just like with music, we need to be cognizant of what claims these outlets are making. The MPA (Motion Picture Association of America) is the organization most adamantly spewing rhetoric about the amount of monetary losses from piracy. The MPA is made up of Netflix, Paramount Pictures, Sony Pictures, Universal Pictures, Walt Disney Studios, and Warner Bros. No conflict of interest there at all! So, just keep my point in mind: one illegal download does not mean one lost sale!

SOFTWARE AND GAMING

Now, let's again flash back to the average 1980s household, where everyone is just having a ball enjoying Tom Selleck driving around Oahu in his Rosso Corsa Ferrari 308 GTS Quattrovalvole in the hit new show Magnum P.I. These every-day people, sitting at home, have absolutely zero knowledge of computing or programming, are miraculously gifted the next big thing—the Apple Macintosh 128K. Now, magically, everyone can own and use a computer. These machines are no longer relegated to the geeks in the know, who make it their hobby to be into anything tech. Steve Jobs has single-handedly democratized computing. And now, with this leap ahead, everyone is also able to copy software and video games to and from floppy disks with immense ease and speed. Well, that can't be good for the profits.

Technically, software piracy has been around as far as the earliest computer programs. People always found a way to make copies and share these with their friends. Of course, businesses that used computers purchased their software, and the people who make these illegal copies were few and far between. They were the enthusiasts and only existed in small niche groups. So, no real problem for Mr. Law and Mr. Software Producer. However, as soon as a young Steve Jobs climbed onto that stage in 1984, and said to the audience "Today for the first time ever, I'd like to let the Macintosh speak for itself", the world changed.

Starting with simple disk copying and evolving into mass online sharing methods, now both Mr. Law and Mr. Software Producer had big problems to worry about. As the number of users began growing due to the launch of Apple's new computer, and the likes of competitors such as Commodore, naturally the software market also began growing. And of course, more software means more theft. From simple "here's a floppy with a game or software on it", to more complicated mail schemes (where users could trade, sell, or request counterfeit copied software), the methods of piracy grew and grew. To prevent this, software companies started adding encryption and keys to their applications. But as always, it's a cat-and-mouse game, and in this case, the pirates were the cats. With the invention of the internet, we got bulletin boards, where not only would people upload software (along with their requisite keys), but they would also just take the guesswork out of it entirely, and crack the software to eliminate any protection whatsoever. These were known as "warez" [20].

With the rise of Napster et al., just like with music and videos, software became even more commonly distributed. BitTorrent made this even easier and faster (the increase in internet speed meant that no one cared that the software files grew bigger and bigger, and it certainly didn't hinder downloads), and thus, the software industry really started to struggle. So now, many major software producers have shifted to online subscription-based and cloud-based models, like Adobe's Creative Cloud and Microsoft Office 365. Generally, this made the traditional forms of piracy far less effective. However, software piracy still exists, just take a quick look at YouTube tutorials. Methods are getting more and more sophisticated, and basically, in a nutshell, are just bypassing application activation and licensing checks. Compounding this with the rise of mobile computing and all these app stores now introduced a new form of piracy, particularly targeting the unauthorized distribution of paid mobile apps. Sure, many companies entirely avoid the risk of piracy by offering their applications free of charge, making money simply through advertising revenues, but that makes the end user experience less appealing. So, in the end, there is no winning, regardless of what the software developers do. This will have to be thought up in the future, though, if I could tell you the solution to this problem, I sure as hell wouldn't be writing this book (probably would be taking my Gulfstream to Bali instead to lay on a beach and sip Mai Tais).

So why is software piracy such a big deal? Well, in general, while both the music/video side and software side have similar characteristics, such as high production costs with low reproduction costs (meaning it's not the physical production that costs money, but the creation of the whole damn thing), the difference is that music and video are "experience" goods [21]. They're viewed in an entirely different light by customers because they buy them (or in our case torrent them) for entertainment purposes. Ain't nobody buying Microsoft OneNote for fun. Now this leaves us with an even bigger issue... no one "needs" to download music or videos. We do it just for fun, so if we don't want to, we can just say "nah, I'm good" and skip it. On the other hand, we can't really do that with software. If you NEED Adobe Photoshop for work, then you NEED it. It is not a fleeting whim; your livelihood may depend on it. And going back to our good friend Manoj in India, if he gets a new job as a photographer (no more slingin' tea on the street corner in Lucknow!), the ability to get Photoshop could be literally (or perhaps maybe it is just figuratively, but you get my point) life or death for him.

The price of photoshop currently in India is over $22 per month... and let me remind you that the average income per month is only $166. The income-to-price ratios of many of these softwares are beyond prohibitive in developing countries. And as these are often NEEDED for earning a living. You can guess what happens next. Yes, Manoj does not need the new Taylor Swift album, but he needs Photoshop. Not that I'm condoning, or endorsing his actions, I'm just proving a point... Further, the music, movie, and TV industries have found ways to battle piracy (through the tons of online streaming services) and this has worked "relatively" well so far. However, the software industry hasn't found such a golden egg yet.

Now, we can't talk about software piracy without mentioning video games. You thought I forgot about them, didn't you? Well, the reason I bring them up now is because for a long time video games were just traditional software. The way to copy, download, or share them did not differ in any way to ordinary software. But there are some fundamental differences. The big difference is your investment into the system. What's another $70 when you already spent $600 on the console itself (or even more so if you spend $2500 building your own gaming PC)? Especially for modern consoles, game manufacturers are really pushing online game sales. Better deals, tons of features, and accessible at your fingertips on your console without leaving the house means more legit purchases. Quick and easy. This logic of "add-ons" also helps gamers rationalize purchases and invest more into their ecosystem. And the more you invest into your ecosystem, the more you get locked in, thus meaning you are going to spend even more! Just like with your Gillette Fusion 5 razor, where you got a bonkers deal for a razor and a free blade for $2.99, now you're spending $59.99 for eight blade refills, because "I already have the razor, what else am I supposed to do?" What a coincidence, it's almost as if Gillette knew you were going to do this all along. God bless human psychology.

Now, with physical game cartridge systems like the Nintendo Switch, it's a little bit more difficult to copy things. Just recently companies started making a workaround that allows you to put a MicroSD card in a makeshift fake cartridge, and loaded with a game, it would behave like a real Switch game. The idea being that you could copy a game ROM from a physical real game (a ROM is what we call the software portion of a video game), and install it to this MicroSD card, which you can then plug into your Switch and play. Too much work if you ask me, probably why Switch piracy hasn't really caught on. What has caught on is the world of emulators and mimicking older games on a PC or other device. You see, older games are smaller and easier to copy, hack, and even play, as they require much less processing power than modern-day behemoths. There are tons of websites where you can download these games and tons of programs that let you play them on your PC. This has gotten so proliferate, in fact, that on Amazon (for just a few bucks) you can even buy little gaming boxes that come preloaded with hundreds of games, so all you have to do is plug it into your TV at home. How nice, Uncle Jeff helping do his part against the big bad software industry. And so publicly too!

WHAT COULD GO WRONG?

So now, you're sitting here wondering "Gee, Marton, everything you said sounds so nice, why wouldn't I stop buying music/ videos/software entirely, and just download everything for free online?" Well, it's not that easy. Let's quickly discuss what could go wrong!

First, let's think of how all of these people making your precious movies, TV shows, music, software, and games are going to end up. Obviously, we have to mention lost revenue. Sure, the numbers the big companies tell us are skewed, of course, they are, they're trying to scare the population straight. But yes, there are of course lost sales, sure not every single illegal download is a lost sale (shoutout to my man Manoj from way back in the previous pages), but some still are. And what do lost sales lead to? Lower revenues, and lower profits all around. No one wants to lose money, and if you do, chances are you're not going to be jumping on that horse again, just to be kicked in the gut a second time. So, companies could invest less in new products if piracy rates are high. This, in turn, could mean less new music, less movies, less shows, less games, and less software [22]. If there is less investment in these areas, less people will voluntarily go into these fields, meaning less innovation as a whole. Because now, Sarah Jane from Boise Idaho doesn't know if she could make a living being a country music singer-songwriter. Or maybe if they were just fledglings starting out in today's world, little Wes Anderson would choose to go into finance instead, because it's clear he couldn't make money on films, so no Grand Budapest Hotel for you. If Kanzuri Yamuchi became a teacher, we wouldn't have the Gran Turismo franchise. Kanye West may choose to go into politics instead, hello Mr. President, no College Dropout album for us… actually, that one might not be so farfetched…

Then we run into the cybersecurity-related threats. How do you know that Daft Punk album you are downloading in that RandomAccessMemories.rar file is actually going to unpack to be a list of songs, and not an .exe file? Frankly, a lot of times, you don't know what you are downloading, and that should scare you. With malware files so easy to hide these days and piggyback onto seemingly innocent stuff, it's a jungle out there. Here is a quick table from one of the leading cyberattack response companies that tells you what's in store for you [23] (Table 1.1).

So, each one of these are just one more thing to be afraid of if you're downloading illegally on the internet. As I said, not as safe as you thought it may be. Just to put things into perspective, nearly half of people in the UK that have partook in stream-ripping say that they or someone they know have been a victim of scams, ID theft, fraud or data loss as a result [24]. Sure, it's

TABLE 1.1

Crowdstrike's Types of Malware [23]

Type	What Does It Do?	Real-World Example
Ransomware	Disables victim's access to data until ransom is paid	RYUK
Fileless Malware	Makes changes to files that are native to the OS	Astaroth
Spyware	Collects user activity data without their knowledge	DarkHotel
Adware	Serves unwanted advertisements	Fireball
Trojans	Disguises itself as desirable code	Emotet
Worms	Spreads through a network by replicating itself	Stuxnet
Rootkits	Gives hackers remote control of a victim's device	Zacinlo
Keyloggers	Monitors users' keystrokes	Olympic Vision
Bots	Launches a broad flood of attacks	Echobot
Mobile Malware	Infects mobile devices	Triada
Wiper Malware	Erases user data beyond recoverability	WhisperGate

only a correlation and not causation, but the odds are fairly clear. Now, what you have to ask yourself, dear reader, is whether it is worth getting that illegal download of The Hunger Games, if it means that all of the personal information on your computer will now be in the hands of a Belorussian hacker, who will either make it all publicly available or force you to pay him $1000 in his choice of cryptocurrency.

Lastly, let's discuss the legal implications of this whole she-bang. We know it. Illegal downloading is illegal. It's in the name.

Yes, you can use VPNs to mask your identity. Sure, you can refrain from uploading your share of files to prevent your ISP (Internet Service Provider) from noticing a large amount of traffic going the wrong way on your IP address. The reason I say wrong way is, as the world moves toward a more and more online environment, downloads are becoming more and more ubiquitous. Everyone HAS to use data downloads for their day-to-day lives, so ISPs can no longer pinpoint pirates from afar, simply by having larger than usual amounts of data coming down. What they CAN do, is look at uploads. Generally, it is much easier to pinpoint anomalies in data being uploaded. Anyways, these are simple solutions, and while being marginally effective, still don't change the fact that you are downloading illegally. And the ramifications of this could be severe. Just take a look at Table 1.2.

TABLE 1.2
Example Punishments for Illegally Downloading by Country [25]

Country	Type of Offense	Penalty
US	Civil (but can become Criminal)	Between $200 to $150,000 for each work infringed, plus legal fees
UK	Criminal	Up to 10 years in prison and an unlimited fine
Netherlands	Criminal	Up to 4 years in prison and a fine of € 45,000
Switzerland	Criminal	Up to 3 years jail and a fine
Russia	Criminal	Up to 6 years in jail and up to 500,000 RUB fine
Sweden	Criminal	Up to 2 years in prison and a fine
Spain	Criminal	Up to 6 years in prison

And there are always articles depicting these sad lost souls, who just didn't know what they were doing, and they simply didn't know that illegal downloading was a crime (cue puppy dog eyes) [26–29]. It's always the big bad record companies, or the evil movie studios that take advantage of poor innocent people. As you may have heard, ignorance of the law is no excuse for breaking it. In fact, Sony has been taking up their own fight against pirates [30]. It all started with them uploading fake torrents of their movies that didn't actually do anything. The whole purpose was to just be super resource intensive to download, and essentially crash the user's computer. They have since moved up a notch, and are using a slew of computers in Asia to launch attacks and take down people that are sharing their copyrighted content.

So, ultimately, it's up to you what you choose to do with all of the above information. Whether you are saddened by the potential loss of the next great artist (imagine all the Swifties without Her Highness Taylor, because she decided to become a bank manager in Pennsylvania), simply scared of the potential technical consequences, or just want to be on the right side of the legal track, whatever it is that you end up doing, at least (having read this) you will now do in a slightly more intelligent and researched manner. As the Dalai Lama once said, "Know the rules well, so you can break them effectively".

REFERENCES

[1] "Did a teenage Mozart really transcribe Allegri's Miserere, after hearing it once in the Vatican?," Classic FM, Jan. 9, 2024. [Online]. Available: https://www.classicfm.com/composers/mozart/transcribed-allegri-miserere-sistine-chapel/. [Accessed: Jun. 19, 2024].

[2] "Online piracy," Wikipedia. [Online]. Available: https://en.wikipedia.org/wiki/Online_piracy. [Accessed: Jun. 19, 2024].

[3] "History of music piracy – A brief explanation," Bytescare, Jan. 15, 2024. [Online]. Available: https://bytescare.com/blog/history-of-music-piracy. [Accessed: Jun. 19, 2024].

[4] A. S. Cummings, *Democracy of Sound: Music Piracy and the Remaking of American Copyright in the Twentieth Century*. Oxford, UK: Oxford University Press, 2013, p. 202.

[5] "Napster," Wikipedia. [Online]. Available: https://en.wikipedia.org/wiki/Napster. [Accessed: Jun. 19, 2024].

[6] D. Kravets, "Napster trial ends seven years later, defining online sharing along the way," Wired, Aug. 31, 2007. [Online]. Available: https://www.wired.com/2007/08/napster-trial-e/. [Accessed: Jun. 19, 2024].

[7] J. F. Buford, H. Yu, and E. K. Lua, *P2P Networking and Applications*, 1st ed. Burlington, MA: Elsevier/Morgan Kaufmann, 2009.

[8] "The death spiral of Napster begins," History.com, Mar. 5, 2024. [Online]. Available: https://www.history.com/this-day-in-history/the-death-spiral-of-napster-begins. [Accessed: Jun. 19, 2024].

[9] Paratii, "A brief history of P2P content distribution in 10 major steps," Medium, Oct. 25, 2017. [Online]. Available: https://medium.com/paratii/a-brief-history-of-p2p-content-distribution-in-10-major-steps-6d6733d25122. [Accessed: Jun. 19, 2024].

[10] M. Stassen, "Music piracy plummeted in the past 5 years, but in 2021 it slowly started growing again," Music Business Worldwide, Feb. 3, 2022. [Online]. Available: https://www.musicbusinessworldwide.com/music-piracy-plummeted-in-the-past-5-years-but-in-2021-it-slowly-started-growing-again/. [Accessed: Jun. 19, 2024].

[11] H. McIntyre, "What exactly is stream ripping? The new way people are stealing music," Forbes, Aug. 11, 2017. [Online]. Available: https://www.forbes.com/sites/hughmcintyre/2017/08/11/what-exactly-is-stream-ripping-the-new-way-people-are-stealing-music/?sh=2fd579fc1956. [Accessed: Jun. 19, 2024].

[12] "Piracy was never killed by streaming," Muso. [Online]. Available: https://www.muso.com/magazine/piracy-was-never-killed-by-streaming. [Accessed: Jun. 19, 2024].

[13] M. Cyrill, "India's per capita income doubles since 2014-15, but wealth unevenly spread," India Briefing, Mar. 7, 2023. [Online]. Available: https://www.india-briefing.com/news/indias-per-capita-income-doubles-since-2014-15-but-wealth-unevenly-spread-27325.html. [Accessed: Jun. 19, 2024].

[14] "How much is Apple Music?," FreeYourMusic, Aug. 2, 2023. [Online]. Available: https://freeyourmusic.com/blog/how-much-is-apple-music. [Accessed: Jun. 19, 2024].

[15] "The difference between VHS and Betamax tapes and how VHS became the household tape," Capture, Apr. 26, 2023. [Online]. Available: https://www.capture.com/blogs/video/vhs-vs-betamax. [Accessed: Jun. 19, 2024].

[16] "Sony Corp. of America v. Universal City Studios, Inc., 464 U.S. 417 (1984)," Justia, Jan. 17, 1984. [Online]. Available: https://supreme.justia.com/cases/federal/us/464/417/. [Accessed: Jun. 19, 2024].

[17] A. Ghosh, "How video and music piracy gave Bollywood nightmares in the 1980s," Frontline, May 10, 2023. [Online]. Available: https://frontline.thehindu.com/books/how-video-and-music-piracy-gave-bollywood-nightmares-in-the-1980s-book-excerpt-when-ardh-satya-met-himmatwala-the-many-lives-of-1980s-bombay-cinema-by-avijit-ghosh/article66830219.ece. [Accessed: Jun. 19, 2024].

[18] E. Van der Sar, "Video piracy visits rose to 141 billion in 2023, report shows," TorrentFreak, Jan. 9, 2024. [Online]. Available: https://torrentfreak.com/video-piracy-visits-rose-to-141-billion-in-2023-report-shows-240109/. [Accessed: Jun. 19, 2024].

[19] F. Duarte, "Countries with the Highest Number of Internet Users (2024)," Exploding Topics, May 7, 2024. [Online]. Available: https://explodingtopics.com/blog/countries-internet-users. [Accessed: Jun. 19, 2024].

[20] "History of software piracy," Bytescare, Jan. 15, 2024. [Online]. Available: https://bytescare.com/blog/history-of-software-piracy. [Accessed: Jun. 19, 2024].

[21] M. Gergely, *Finders, Keepers; Stealers, Reapers: An Experimental Examination of Three Antecedents of Software Piracy Behavior*. Ann Arbor, MI: ProQuest LLC, 2015.

[22] M. Gergely, "A system archetype analysis of digital music piracy," in *AMCIS 2017 Proceedings*, 2017, p. 58. [Online]. Available: https://aisel.aisnet.org/amcis2017/TREOs/Presentations/58. [Accessed: Jun. 19, 2024].

[23] K. Baker, "The 12 most common types of malware," CrowdStrike, Feb. 28, 2023. [Online]. Available: https://www.crowdstrike.com/cybersecurity-101/malware/types-of-malware/. [Accessed: Jun. 19, 2024].

[24] "Dangers of illegal streaming," FACT, Mar. 29, 2023. [Online].
     Available: https://www.fact-uk.org.uk/consumer-advice/dangers-
     of-illegal-streaming/. [Accessed: Jun. 19, 2024].
[25] D. Crawford, "Is torrenting illegal?," ProPrivacy, May 4,
     2022. [Online]. Available: https://proprivacy.com/vpn/guides/
     torrenting-illegal. [Accessed: Jun. 19, 2024].
[26] E. Jackson, "Nobody gets sued for illegally downloading mov-
     ies, right? – Think again, Canada" Financial Post, Feb. 15, 2019.
     [Online]. Available: https://financialpost.com/telecom/media/
     massive-infringement-movie-rights-holders-are-suing-illegal-
     downloaders-and-winning. [Accessed: Jun. 19, 2024].
[27] S. O'Shea, "Lawsuits for movie downloading and upload-
     ing," Global News, Feb. 9, 2019. [Online]. Available: https://
     globalnews.ca/news/4933339/lawsuits-movie-downloading-
     uploading/. [Accessed: Jun. 19, 2024].
[28] A. Holpuch, "Minnesota woman must pay $220,000 for 24 songs
     illegally downloaded," The Guardian, Sep. 11, 2012. [Online].
     Available: https://www.theguardian.com/technology/2012/sep/
     11/minnesota-woman-songs-illegally-downloaded. [Accessed:
     Jun. 19, 2024].
[29] M. Hardigree, "How the RIAA took my vintage Mustang,"
     Jalopnik, Oct. 19, 2010. [Online]. Available: https://jalopnik.
     com/how-the-riaa-took-my-vintage-mustang-5667666.
     [Accessed: Jun. 19, 2024].
[30] L. H. Newman, "Report: Sony pictures is using its own cyber-
     attacks to keep leaked files from spreading," Slate, Dec. 13, 2014.
     [Online]. Available: https://slate.com/technology/2014/12/sony-
     pictures-is-using-ddos-attacks-to-keep-its-leaked-files-from-
     spreading.html. [Accessed: Jun. 19, 2024].

2 If You Teach a Man to Phish… – The Mechanics of Internet Scams

When I was younger, I used to visit town fairs with my family. They only really came once a year, and at a time when school was out, so it was meant to be something special to celebrate. Rigged games, greasy food, neon-colored diabetes-inducing sweets, and a random pony with a rainbow mane and foam horn hanging off its head. Off toward the end of the fairground, there was always that one tent where an older lady sat with shaggy robes and long, mangled gray hair. People were excitedly lining up to speak with her, but when I asked my mom why, she told me, "we don't deal with that stuff". I asked her what stuff she was on about, and she told me that the woman claimed to be a fortune teller and did palm reading. She proceeded to tell me that this old lady was just scamming people, and that her readings didn't mean anything. However, people just kept lining up, giving this woman money just for fake readings.

I just found it so interesting as a kid. It's not a matter of whether or not I believe in their powers or spirits or any cosmic entities, but it had just seemed so suspicious at the time. If she were a seer, where were her tools? Her incense? Her candles? Her charms? She was literally just an old lady in a tent that needed a haircut. Observing her more while waiting for my parents to take my siblings to the bathroom, she wasn't even properly looking at her customers' palms. "Open up, mmhhhmm,

DOI: 10.1201/9781032679389-3

mmhhhmm, you will earn great rewards from your endeavors".
What endeavors would she be talking about? She probably didn't
even know and was just keeping it vague to sound mysterious.
Her customers, of course, would eat it up (like those deep-fried
Snickers bars they also used to sell at the very same fair) because
they have something they're struggling with, and everyone wants
to hear that they'll pull through with great success. I can see how
this lady hooked them, appealing to victims with silent demons
peering out at the world from behind their eyes. Wouldn't you be
tempted to hear about how those demons will disappear someday?

Scams like this have been around for ages. The oracle scam is
one of the oldest, dating back to at least the 8th century BC when
the Ancient Greek Delphic Oracle rose to fame [1]. Scammers
would pretend to be one of Delphi's oracles and offer blessings
or prophecies (made up on the spot, of course) to people who
would trade their money or valuables. People were a lot less
skeptical of fortune tellers back then (don't forget, for a time,
Mediterranean cultures revolved around mysticism). Fast forward to the 20th century AD, where this type of scam takes on
an interesting twist; a member of the Nigerian royal family is
looking for a good Samaritan to lend them some money to
resolve whatever issue they are having, promising to reward
their hero once all is well again. Although it was slightly different (helping someone for a bigger reward in return, rather than
receiving a fortune or blessing), the premise was the same (giving your money to someone expecting something back, and
likely receiving nothing). Fast forward once more to today, and
we still have people losing their savings to criminals claiming
they'll repay whatever help they receive or offering non-existent
or false services (but now with artificial intelligence also mixed
in). If this game has been played for centuries, how are people
still falling for its tricks? As Kevin Mitnick once said, "Those
evil hackers are always looking for the weakest link in the security chain. In my opinion, most often these are people, not technology" [2].

Mitnick was a very well-known social engineer who evolved to become a big guy in the cybersecurity world. In case you haven't heard this term before, a social engineer is someone who coerces their target into revealing information or performing a specific task by charismatically pretending to be someone of authority [3]. This charisma is enough to convince the victim to give stuff up, for some people anyway. Think of it like this: If someone called you and said they were an IT support technician from the company you work for and claimed that they discovered a server conflict with your account, wouldn't you believe them? You can't risk having any problems that could potentially impede the progress of your work. This is one way that social engineers can try to trick you into giving up your account information – by causing minor panic to get you to comply. THEN, later on, you remember that IT people wouldn't (shouldn't) be asking for your credentials directly, but by then, it would be too late. So, this was how Mitnick put himself on the FBI's Most Wanted list between the 1980s and 1990s: by social engineering his way into companies – unauthorized, of course. He didn't do it for money, he just wanted to see if he could. Nonetheless, a crime is a crime, and he was eventually sentenced to jail in 1995, where he stayed for five years to think about what he's done. After his release, he went on to be a prominent figure in the cybersecurity world, starting a security consultancy company bearing his name (Mitnick Security) and writing several books about his personal experiences as a social engineer, among stories of others following his way of life. In a matter of speaking, he was the perfect person to do all this; he was on the other side for a while and has seen how criminals would approach the idea of compromising a company, especially through their employees. Which, again (I should reiterate), Mitnick did just for the hell of it, not to actually gain anything except for the satisfaction of setting his mind to something and succeeded with it (for most others that he shared a jail cell with this was not enough - they were much greedier). He wrote four books in total: The Art of

Deception, The Art of Invisibility, The Art of Intrusion, and Ghost in the Wires. All are insightful on how humans can be taken advantage of to gain unauthorized access to systems, as well as other ways infrastructures are flawed and how we can protect ourselves.

Mitnick's tactics, as described in his books, often involved using a phone to pull off his schemes; he would call around pretending to be several different people to get the information he wanted, so that he could gain unauthorized access to his targets. When he was younger, the internet was not accessible to everyone, so most of his tactics had to be conducted over the phone. You know those prank calls where someone would ask if your refrigerator was running and then snicker and tell you to go catch it if you said "yes?" Of course, those were just prank calls, but Mitnick (and others who followed suit) did this for a different reason: to gain information. "Pranking" others for sensitive information over the phone at that time was called "**phreaking**", and essentially was the experimentation and exploitation of the limitations of telecommunications systems (obviously, as the name suggests, this word came from the word phone) [4, 5]. As the internet became more accessible, these ploys evolved from purely occurring on the phone to also happening through email in what became known as "**phishing**".

CASTING THE LINE

To help you imagine how phishing works at a higher level, think of actual fishing (the correct spelling, this time). A casual fisher (or fisherman to you and me, but bear with me, I'm making a point here) would just grab a pole, maybe some bait if they're bothered enough, and go out to the nearest lake and cast their line, hoping to catch something. They might, they might not. It depends on the environment, the bait (if any), the weather, the time of day... lots of factors. A more seasoned fisher, on the other hand, will do a little bit more research. They might look into the season and if it's even worth going out, the best locations at the time, the kind of fish they could catch at any given location, what kind of bait they should buy for that specific kind of fish, and the best time of day to go. They would put a lot more effort into their fishing endeavors and likely, as such, have a higher rate of success. Phishers (there's my point, now we're talking about cyberspace) work in much the same way. A casual phisher would try to put down simple bait in hopes that someone is stupid enough to fall into their trap. A more intricate phisher would study the environment they plan to place their trap; the type of users they might encounter, potential interests of those users, specially and carefully tailored bait (we'll get to what this would look like later)... Obviously, users would be more likely to fall for the latter phisher's traps than the former's. The word "phishing" is quite literally derived from fishing, just replacing the "f" with "ph" to follow the "phone phreak" trend.

The term "phishing" was coined in 1996, when a cracking (breaking through license limitations) software named AOHell mentioned it throughout the application text [5]. It involved an AOL (America Online – the top internet provider at the time) scam where deviants would phish for the account information of legitimate users, who would become their victims (bless their hearts), to get free internet access. After the year 2000, phishing attacks evolved to focus more on online banking, web-based marketplaces, payment systems, and social media. Instead of attempting to gather credentials over the phone, phishers would register numerous domains that masqueraded as legitimate web-sites to trick them into submitting their user information [5, 6], such as "amazone.com" instead of "amazon.com". Email pro-grams were created to automatically send emails to people with these "fake" links leading users to the malicious domains. And boy did it take off... Between the years 2004 and 2005, 1.2 mil-lion users in the US had become victims of phishing attacks, resulting in $929 million in losses [6]. And don't forget, at that time, the internet was a whole new world for most people, and they didn't understand that "don't talk to strangers" also applied to online interactions. So, you would think the people would know better by now… but I'm getting ahead of myself. Phishing attacks continued to evolve, either by becoming more intricate, or by combining them with other attack methods, such as malware injection. Phishing turned into a whole umbrella of attacks, rather than just individual scam we set out talking about.

There is no sole way to target the potential victims of phishing-related scams. These attacks can be broad and untargeted, just setting out some bait somewhere and hoping some poor user would bite. But scammers do also target specific users, and that is called **spear phishing** or **whale phishing** (also known as **whaling**). So once again, plain phishing is broad. Any fish (victim), and size, any bait. Spear and whale phishing are targeted, with the specific difference between the two depending on the "size" of the target. I might use a plain ol' spear to try and catch a barracuda (using specific tactics to catch a specific fish or type of fish). In this case, maybe I need to target a secretary to gain access to an internal system, ergo, that's my targeted fish. Or maybe I'm looking to steal unpublished manuscripts of great books (like the one you're currently reading) from academic authors, so I target those pretending to be a big publishing house. In whale phishing (fishing analogy continues) I can equip myself with super ridiculously specific tools, like harpoons and a giant boat, if I'm aiming for a whale (generally equating to targeting a higher-up executive within a company, like a big manager or a C-something-O). And for the record, I don't condone whale fishing (who does, but it needs to be said), I just wanted to give you an idea of the enormity of the target. Anyone can become a target of phishing, if they are not careful enough, which is what happened with Snapchat. This is a company that has a fairly popular app (you may have heard of it) where you can share ten-second media clips with people (privately or publicly) and they disappear for good once viewed, or after twenty-four hours when the post expires. Snapchat fell victim to such a whaling attack in 2016, when an HR employee responded to an email (supposedly) sent by their CEO. The employee responded to the email, attaching the requested information... all of the payroll data for the entire company [7]. Unfortunately, it was not, in fact, their boss that they sent the data to, but the phisher. They can make media disappear after ten seconds, but they can't do that to mistakes like this. But they're not the only ones that fell for this kind of scheme. In that very same year, an executive at Seagate responded to a whaling email, and disclosed the financial information for all current and former employees, causing a breach of income tax data for about 10,000 people. The same thing happened over at Mattel, where their senior financial officer wired $3 million to a scammer, because he believed the request came from the CEO [8].

But targeted attacks like those above aren't the only ones we have to keep an eye out for. We have attacks that are targeted, but on a larger scale too! Hooray! These aren't targeting specific individuals anymore but have upped their game to targeting entire groups (those with something specific in common) at once. So, in the wild, a watering hole is a spot where all the animals gather to drink water. It's treated as a sanctuary where predator and prey have a common interest: sustenance. To the malicious mind, one might think that this spot would be the perfect opportunity for the predators to strike; the prey would have their guards down as they drank water, and there would be plenty of options to hunt down. Or, if it's not a predatory animal that might want to take advantage of the group, an insanely evil person might see this as an opportunity to poison the water and kill all the living beings that consume it. In a way, humans are similar to animals in the sense that we have places where we gather for sustenance or interact. In the workplace, this might be the water cooler in the kitchen, hence the term "water cooler chat", where colleagues stand around to converse as they drink their... well, water. Should someone poison the water cooler... well, you'd kill the whole company (cybercriminals here would have a lightbulb moment). And yes, this kind of behavior happens on the internet. Scammers come by and "poison" a website, and then the users of that website, around the world, are affected, as they all relied on using that specific service.

This is how a **watering hole** phishing attack works: attackers discover the websites that their target group of users most frequently visits and inject malicious code or create a clone of the site for legitimate users to mistakenly access [9]. One example of this type of attack happened in 2019 and was named Holy Water, which targeted religious websites, charities, and volunteer programs [10]. Anyone visiting these kinds of websites became a target, whether they were kids trying to learn about their faith and get on God's good side, or extremists shoving their propaganda in others' faces through their monitors. This was a broad phishing attack, but it only affected users that visited those specific websites. Generally, there are two ways a Watering Hole attack can be performed, either by taking advantage of vulnerabilities found on legitimate websites, or by cloning or spoofing those sites, and registering a domain name that looks very similar to the real one (like what I mentioned before with amazone.com). Online shopping websites tend to be the disguises for the latter method, such as eBay or Amazon. Attackers register an email and website domain that looks eerily similar to the actual URL (link addresses) and BAM! You get that email from amazons@amazons.email asking you to verify your payment method [11]. So, going back to the Holy Water attack… what exactly did that one do? Well, the phishers injected malicious JavaScript code into these specific religious websites and gathered information about all of their users. All it took was for you to click on one affected link, and the scammers had access to your info. Fun stuff.

Okay, so… Phishers have been relying on emails as the primary medium for their attack for decades now, and they have started to evolve. They've realized that they can't always rely on emails for their attempts. Well, ladies and gentlemen, they've looked back at the ancient ways of the phone phreak and started using telephone calls as their mode of attack. We have come in full circle. Now called "**vishing**", or voice phishing, phishers abuse the telephone (or Skype or WhatsApp calls, etc.) to call their victims and attempt to persuade them into complying with demands and giving up sensitive information [12]. Essentially just phone phreaking all over again. As old as this technique is, it apparently still works though! You know what they say, "if it ain't broke, don't fix it". And meanwhile, people have forgotten that phone calls can be used as scams. Back in 2019, after the UK Parliament had already been bombarded with other phishing attempts (that mostly failed), attackers resorted to vishing members instead, attempting to trick them into giving up their credentials over the phone [13]. And guess what? It worked! The one (kind of) caveat here is that my fellow millennials and I hate phone calls. We find them uncomfortable, awkward, and might even ignore them most of the time (even the important ones – sorry Marton!), but we do like our text messages. If phishers don't find success in vishing, they might convert to "**smishing**", and that's where text messages (SMS or through messaging apps like WhatsApp or iMessage) are used as the medium for attack. These phishing messages are sent to targets through text, instead of email or phone calls, and claim to need "urgent" attention [9]. Banks, retail stores, tech organizations, and delivery services are often used as disguises within the messages to convince the targets to comply, such as US Bank, Amazon, Apple, or American Express [14]. After all, you'd hate to miss out on your Amazon delivery just because you didn't send a confirmation message, right? A similar case can be seen in Figure 2.1, which happened

Sunday 10:39 AM

Emirates Post: Your parcel has been stopped from delivery, please update the address as soon as possible to pay for shipping, reply 1 to get the link: https://dbnl1.top

FIGURE 2.1   A smishing attempt from a phisher pretending to be Emirates Post.

to a cousin of mine in 2023, where an alleged delivery service claimed that the address where a package is supposed to be delivered is missing and the user must follow a link to resolve this issue.

An SMS smishing attempt message stating, "Emirates Post: Your parcel has been stopped from delivery, please update the address as soon as possible to pay for shipping, reply 1 to get the link:" followed by an inactive, malicious website address: "h t t p s, colon, double slash, d b n 1 1 dot top

One quick look at this message should be enough to convince you that all of this is fake. I mean, Emirates Post does exist – it's the governmental mail and delivery service of UAE. But the wording just sounds wrong, and the link should definitely give away the fact that this is fake. The sender is just trying to get information from the target. Sure, it says, "please reply 1 ..." and you can always just never click the link that's sent back, but the phisher still gains something here: the fact that the contact number they phished is active and gullible. If you don't respond to this demand, they can always try again with a different scam until you finally give in.

These types of phishing attacks are relatively easy to pull off, but they are not the sole form of online scamming. Don't forget, phishing is mostly just getting information or getting someone to comply to a demand, but the attacker wouldn't gain anything immediately. It all takes time. If you want immediate results, there are alternative options. For instance, the Nigerian prince example that I gave at the beginning of this chapter. Our prince would try to trick the victim into sending money to the attacker by convincing them that they were indeed rich, just in a tight spot at that time, and they would pay them back later. This is a scam that dates back to 1920, where a Professor Crentsil (what kind of professor you may be wondering, why a "Professor of Wonders", of course) attempted to contact people, in what is now Ghana, claiming to have magical powers and saying that he could benefit his correspondent if he were loaned just a small sum of money [15]. Within one year, Crentsil was charged on three counts of what is known as "419", an article number within Nigeria's criminal code referencing fraud. Ever since, the Nigerian prince scam is synonymous to **419 fraud**, which is a type of **advance-fee** (or sometimes called **beneficiary**) scam, where you pay some money in advance, and hope for more in return later [16]. By now, people might be familiar with every-one's favorite Nigerian royal family, so criminals might resort to using other identities, usually businesspeople or that of a member of a wealthy family. If you've seen Tinder Swindler on Netflix or followed this case on the news, you might be familiar with Simon "Leviev", who scammed numerous women on the dating app Tinder. The reason his last name is in quotes is because it's not his real last name, he only used it online to try and legitimize the claim that he came from the Leviev family, known for Lev Avnerovich Leviev, a diamond tycoon. Women he matched with would believe him and a relationship would blossom. Once trust had been established, Simon would ask for money, claiming that he is in some dire situation and needs some cash to get out. Except he'd be perfectly safe and would instead use the money that his first girlfriend sent him to woo a second woman. Then, he would repeat the cycle: ask for money from the second woman to target a third, and so on, and so on. Wow, both a 419 fraud AND a sort-of Ponzi scheme (an investment scam that pays early investors with money taken from later investors to create the illusion of big profits... but in this case it was reversed as Simon took the money from the early "investors" and gave it to the "later" ones)!

But in fact, it was even more than that. The Tinder Swindler was a combination of THREE kinds of scams: the 419 scam that we just talked about, the sort-of Ponzi scheme, and another new one known as **catfishing**, or **romance**, or **dating app** scam. Allow me to set the mood and light some candles to give you a better idea of what this latter one is. A friend of mine met some guy online through one of the video games that we played. The two of them immediately clicked. Soon after, we invited him to join our Teamspeak server (no Discord at the time) and to join our matches. He was genuinely a nice guy, and cute too (he had sent her a picture of himself and, naturally, she showed us). But we could see that our friend was head over heels for him, so we warned her that we don't actually know him beyond our online interactions. Against our judgment, they got into a relationship after only three months of knowing each other. This guy started eating up her life: taking her time, isolating her, needing to always be virtually with her, needing her to always be on-call, demanding status updates whenever she was out... All the signs of an abusive relationship (well some of them, but you get my point). Worse than that, he kept asking her for money. Somehow, he was always short a few bucks for things, like ordering food, needing to pay the bills, paying his therapist (foreshadowing maybe?), gas money... Always with the excuses! We tried warning her again, but she refused to listen, seeing his abusive demands as signs of love and caring. And she needed to return the favor! This went on for over a year before he admitted that he had been lying to our friend about everything from his looks to his intentions, and he *POOF* just eventually disappeared.

It was devastating to see our friend go through all of that, but unfortunately, she's not the first or only victim. Her experience is typically how these romance/catfishing/online dating scams work: the scammer creates a false identity, tries to get to know their target, uses photos of "themselves" and other relevant personal information to build trust and establish a connection, and, once the bait has been taken, they string the victim along for as long as they need, asking for money or any other items [16]. Usually, these might happen on dating sites or applications, but, as with my friend's case, they can happen on other online social platforms too. And one more thing to keep in mind here is that there could be a second victim as well: the person whose photos were used for the scam (Tinder Swindler didn't do this, but my friend's scammer did). These photos are used without the owner's permission and often taken from sites that store scores of images like Google, Instagram, Facebook, or Tinder [17]. And more recently, catfishers are not stealing these images from social media... Nope! They have evolved into taking advantage of generative AI and making brand new images altogether... but I will go into more detail about this in my AI-dedicated chapter.

Let's switch some gears now: what happens when the person you're suspicious about offers to give YOU something? That sounds strange, doesn't it? Usually, if someone wants to scam you, they try to convince YOU to give THEM something, not the other way around. But let me put it this way: you know those warnings about people claiming to be repairmen coming up to your door unannounced and asking you to let them in so they could check if your washing machine was working fine? Take that ploy and try to imagine it from a tech perspective. What could someone possibly offer you? If you guessed device repair, then good job, you get a cookie. To get this to work, though, the bad guys would hope that you know nothing about computers. In a **tech support** or **repair** scam, attackers will pretend they work for a big-name tech company, like Microsoft or Google, and try to convince you that they need to access your device [16]. They'll tell you that you need to download some remote access software, which will allow them to gain direct access to your computer too. Well, you just gave them free access to your device, which means they can directly install any malware that they want, promising that it'll solve your problems. Or they can even snoop around and try to gather sensitive information and files you might have saved on your hard drive. Once they've succeeded in doing all that, they have the power! They can continue scamming you or run off with the information you basically just handed to them and then take advantage of you behind your back.

And of course, there are tons more types of scams other than the ones we talked about above. There is physically not enough space in this chapter to talk about all of them. BUT what you should be aware of is that the one thing that all of these scams will always have in common, is that they will always target your *trust* and *lack of knowledge* (or naivete). With tech support, scammers will hope that you lack enough technical knowledge to trick you into thinking they know more than you and just want to help. Such nice guys... Another way you can be predated is by your potential financial aspirations. Giving money to someone in general tends to be suspicious nowadays, even to friends and family. In a world like today's, I can't even be sure that I'll ever get it back (and I kind of need my own money to survive, you know). But nonetheless, people keep telling me to invest in things (real estate, gold, Bitcoin, Gamestop shares... Anything's become an investment nowadays!), but I've always been a bit hesitant about that. I mean, all these codes and numbers and keeping track of the market... I see my brother doing that at his banking job and it's overwhelming, even as a spectator. So, what can I do? Get a broker! Someone who's more qualified in the field of finance and investments... that's what I need. Someone to do that all of this for me for a small percentage fee. And then I can really get into the market and start earning... Sounds a bit more plausible, and I might be more tempted to take the dive. And it seems that scammers have realized this as well. In **pig butchering** scams, you'll encounter someone that is pretending to be a broker and offer to help you invest in stocks or cryptocurrencies, but then they'll disappear once you give them your money [18]. Of course, you wouldn't give your money to just some random person. First, there needs to be some level of trust. The scammer would want to talk to you, get to know you, and make you feel comfortable enough with them until they finally pop the question: "Will you allow me the honor of helping you invest in things?" And then they seal the deal with a big, bright, shiny, glamorous... investment portfolio. With all of this, you're hooked. "Yes!" You can't help but exclaim and you start to feel all giddy and excited about your fabulous new futures (investment pun here) and all of the new and exciting opportunities that will come your way. The scammer will then gently hold you by the hand (metaphorically speaking) and lead you to a "totally legitimate" investment platform where you can begin "investing" [18]. They might even try to convince you to invest more, until you finally refuse or simply can't afford to. And then when you finally go to withdraw your earnings, you'll likely find that the platform is no longer accessible or available, and neither is your new friend. You're left with nothing. You've been slaughtered, much like a butchered pig.

REELING IN REASONS

I can almost feel your eyebrows cramping from being furrowed and hear the little old cogs in your brain working to comprehend all of this. By now, you're probably wondering how humanity could be so corrupt that people would want to hurt others like this. We're all on this earth together, we're all humans, we're all one people. How can they justify their nefarious actions? Well, there's a little something called **neutralization theory**, where people will come up with reasons to justify their deviance. Let me pick another friend to use as an example (they don't mind... I hope). I'll call her Lydia. Lydia is a heavy smoker and has been since our final years at high school. I don't particularly like that she smokes since I worry for her health in the future, and she has admitted that she regrets ever picking it up because she relies on it as a coping mechanism. If I want to be evil, the next time I see her, I can take her cigarette pack and tear open each of the sticks inside. This isn't exactly a crime I would go to jail for, but she wouldn't exactly be happy with my actions either. I would've technically stolen her property (the box and the sticks inside), which she had spent her hard-earned money on, and "maliciously" destroyed it (in her eyes). Even if she's upset, if I were a bad friend, I wouldn't feel bad about doing this, because I would try to justify *why* I did it, which would be a couple of reasons: first, that I think smoking is bad (my beliefs only, I mean no harm to our nicotine fueled amigos out there) and second, that

I was actually helping, not hurting, my friend (denial of injury). With this, I *neutralized* my "evil" actions, by justifying them. Of course, this is from my own perspective, just to make myself feel better about it. The reality is that my friend would still be fuming. This is how scammers try to think when they commit crimes; they justify their actions through neutralization theory, by: denying responsibility, denying that any harm was done, suggesting that the victim deserved what happened, shifting blame, or claiming that they did something for a higher authority (e.g., the big four – self, family, government, God) [19].

When it comes to these online scams, criminals tend to comply with the rules of neutralization theory to justify why they scammed people in the first place. For example, a study was done where an interview was conducted with two advance-fee scammers that operated from Africa, to learn how they got their expertise and why they committed their crimes [20]. The two scammers discussed in great detail how they got started with pet scams, where they try to give away a (non-existent) animal, because they can no longer keep them for whatever reason (like their landlord having a "no pets allowed" policy). They would claim that they were willing to give the pet away for free due to the urgency of the situation. When the potential victim takes the bait and inquiries about the animal, the scammers would be ready with plenty of pictures and videos of random pets from the internet, as well as fake documents. If the victim was interested, the scammers would magically be "extremely far away" from the victim, but willing to send the pet anyway, if the victim was willing to pay the shipment/transfer fees. For a free pet, this might be a small price to pay (damn, those purebred Tibetan Mastiffs can get really pricey if you want to get one from a reputable place). Except the scammers would now start to create more and more excuses for the victim to send money, like a change in the delivery method due to weather conditions, vaccinations for the animal to be allowed to enter the country, etc., until the person on the other end finally changes their mind and refuses to send more cash. When asked why they scammed their victims (who they kept calling 'the Westerners' throughout the interview, so we have a general idea of who they were targeting), the fraudsters rationalized that other scammers they knew typically had an "extravagant lifestyle" resulting from their successes (something for a higher authority – they wanted a better lifestyle to support themselves and their families) and that it constituted for reparations of the colonialization of their countries (essentially saying that their targets deserved it due to historical events that caused great suffering to their people and country).

Catfishers/Romance scammers aren't too different in their line of thinking. They still follow the rules of neutralization theory, just with different reasoning. In an extensive study on how criminals rationalize romance fraud, researchers found that victim blaming was key in the scammers' arguments. Offenders stated that both parties in a relationship sought to reap the benefits, be it financial (them) or emotional (the victim), so therefore, the victim should just accept their fate [21]. Additionally, the scammers would justify their actions by minimizing the harm done (in their mind) since there were only some monetary losses. They even go so far as to deny ANY responsibility for their actions whatsoever and say there was no victim at all. They stand by these perspectives since catfishers must put quite a bit of effort into their scams, building relationships and sometimes even gifting their targets to keep them interested. ADDITIONALLY, the blame is shifted to two people: the victims of romance scams since they were gullible enough to fall for the scam, and law enforcement agencies for not doing enough to stop such crimes from occurring. In pointing fingers at law enforcement, the scammers are basically admitting that they believe these crimes are low risk and less of a crime than traditional offenses [20–22]. So, even if all these reasons are different than the ones given by the African scammers, the end result is the same (money stolen from victims) and the general justifications are also the same (self, family, government, God).

So apparently, as we have now learned, money makes the world go 'round! When it comes to why online scams are committed, the common denominator seems to be... *drumroll please* ... money. Surprise, surprise! Whether we're looking at any of the crimes I described earlier, or the studies I just mentioned, one of the key motivations in cybercrimes is almost always money-related in some way, either because of poverty, low-income living, or unemployment (on the scammer's side) [20, 22]. Tinder Swindler – money. My friend's scammer – money. 419 scams – money. Pig Butchering – money. Money, money, money. People struggle to live, and they need to make fast bank to survive. If not for cash, then the scammer probably joined as a part of a group; they knew people that were scamming and were inspired to join their group, especially when witnessing their successes of their friends first-hand [20, 22]. "We grew up together. Then he started living a flashy lifestyle, expensive whiskey, girls, nightclubs. I started asking him how he did it. He said he would just talk to people online, people he had never seen, and they would send him money" [20]. These friends would go so far as to cover the expenses of the inter-viewed scammer during outings, adding onto the motivation for a self-sufficient life that they already felt. Whenever Lydia pays for my coffee, I cover her sushi, then she gets me ice cream and I buy her hotpot, and the cycle continues. That's what friends do for each other. There's no reason it should be any different for the scammers. If not for the cash, or the mild peer pressure, there's also the chance that the scammer actually enjoys what they do. To some, they might even consider it a game, which further cements their justification for committing these crimes in the first place [21].

SNAGGED AND SINKING

So, we've looked at the evil side of things, namely the what's, how's, and why's. At least, the 'why' we looked at was why online scammers do what they do. We still haven't asked the age-old question: why do people keep falling for these tricks?

What makes people look at a message that is clearly phish-bait and give their information away so freely? Well, for one thing, humans are social beings; we need to talk to other people, do things that make us feel less alone. Unfortunately, this need for communication can make people susceptible to becoming the victim of a scam. Loneliness and isolation can affect anybody enough to push them to accept the first sign of socialization they encounter, especially when it comes to older people [23, 24]. This happens with romance scams, where people are desperate enough for attention that they become a likely victim (cue Baby Reindeer reference). It's the interaction, sending emotes, sending pictures, liking each other's posts. Wouldn't you feel happy if a person you enjoy talking to starts liking all your content?

And if the person you like to talk to sends you a message, wouldn't you feel tempted to read and respond immediately? What about when they send you links to show you something? If you have a modicum of self-control, you might first check where the link leads to and maybe even question the sender about the content. Unfortunately, some people don't hesitate like that. There seems to be this connection between impulsivity and the likelihood of falling victim to these scams [23,25–27]. If someone clicks every little link and image they see, they might not realize that they're falling into a trap. The other possibility is sending money; if the person they enjoy talking to requests money, an impulsive person might send it, no questions asked. Impulsive people will fail to see the scam right in front of them. And once again, if you hesitate, that tiny period of self-reflection may allow you to realize HEY! IT'S A SCAM! Another big factor in this, of course, is a user's familiarity with computers. As we said before, scammers feed off of your *trust* and *lack of knowledge*. People that present higher digital literacy levels or practice safe media habits tend to recognize disinformation and scams better than those that are less familiar with the internet [26, 27]. This is why we're writing this book! To help YOU improve your digital literacy ever so slightly, so that you become aware of these pitfalls on the internet.

But don't get me wrong, the flip side of this can be an issue as well. And there, the downfall is cockiness. This confidence can also get in the way of ACTUALLY succeeding to tell the difference between scam and normal messages (we call this **optimism bias**). Just ask Icarus. Oh… Wait… You can't. Well, the first reason why you can't is because this is just a story in Greek mythology. The other reason has to do with what happened to Icarus and his father, Daedalus. The two were trapped in the labyrinth of Crete, where a Minotaur (half-man, half-bull monster) lived and made sure that the two never left. Daedalus, being the skilled craftsman that he was, crafted wings for himself and Icarus, so that they could escape by air. Icarus, enjoying his flight a little too much, ignored his father's warnings about going too high. Icarus flew too close to the sun, causing the wax holding his wings together to melt, and he plummeted into the sea. Daedalus, knowing better, continued on to freedom. Being too confident in our ability to do anything hurts us. And while it's okay to feel good about yourself and knowing that you can avoid scams easily, basking in that glow might just melt your wax. There is a link between overconfidence and scam susceptibility; being too confident about your abilities to detect scams makes you pay less attention to whether you are about to get caught up in one [23, 25]. Because, you know, you're supposed to just KNOW, since you're so great at pinpointing fake messages and all (sarcasm).

On the other end of the spectrum, though, we have people that just don't know enough, or people that generally don't possess adequate cognitive abilities to determine whether a message is a scam or not (cognitive abilities refer to a person's memory, critical thinking, processing speed, and attention/perception). These are also things that degrade as we grow older, further affected by disorders that we may develop with age, such as dementia, Alzheimer's, or sensory loss. And the sad truth here is that some people are willing to take advantage of this. Ergo, cognition and age are BOTH factors in a person's ability to detect scams [25–27]. As much as we can preach cyber safety, there's a chance

TABLE 2.1
US Scam Cases and Losses in 2023 [29]

Age Group	Number of Cases	Losses
60 and over	101,068	$3.8 billion
50–59	65,924	$1.7 billion
40–49	84,052	$1.5 billion
30–39	88,138	$1.2 billion
20–29	62,410	$360.7 million
Under 20	18,170	$40.7 million

that the lessons will be forgotten, misunderstood, or just not processed at all. While age factors into potential victimization, is it actually as big as one might think?

Let me broaden the question a little bit: Who exactly DOES fall victim to online these scams we've spent a chapter talking about? We all know the stereotype... "older adults are more vulnerable to scams" [28]. Well, dear friends, it's not that simple. Sure, looking at the age groups, people aged 60 and above are most likely to fall susceptible, that's absolutely true. But by how much? Check this out in Table 2.1.

So, in 2023, seniors above the age of 60 reported the largest number of cases and losses [29]. But the three age groups immediately preceding them weren't that far off! The report numbers are definitely not too radically different, but what's interesting here is that despite the youngest group having the least amount of financial losses, they had the MOST number of reports (so, essentially, this younger group either fall for more scams but only get swept up in them to a lesser extent than their elders, or they report them as they see them). Altogether, 1 in 3 Americans have said to be victims of financial fraud at some point in their lives [29]. That's a pretty heavy hit. To dive a little deeper, like with anything, anywhere, gender differences can also be looked at. Fifty-three percent of women were more likely to accept the fact that they could become the victim of financial-related scams, as opposed to men, out of whom only 45% believed the same could happen to them [30]. And is this cockiness, or does it hold true in reality? Well, a smaller study done on the other side of the pond noticed a similar trend, where 6.7% of women had fallen victim to fraud, as opposed to only 5.8% of men [31]. And liability to scams isn't just discriminated by gender, no, no. Different ethnic groups can have different susceptibilities as well. For instance, in the UK, those of African descent, followed by mixed-race individuals, were the ones most likely to fall for scams [31]. And don't forget, regardless of age, gender, or ethnic group, becoming the victim of a scam can also affect you emotionally and mentally (this can be in the form of fear, anxiety, depression, shame, and anger, in addition to a big hit to your confidence) [31].

So, what types of scams are most of these people falling for? And does this all even matter? Well, let's first take a look at the history of these numbers, just to quickly prove a point. In the US alone, the number of reported scams since 2019 has doubled (to 880,418 in 2023), and the losses QUADRUPLED (to $12.5 Billion in 2023) [29]. So, I'm not just trying to freak you out, this shit is real. There were nearly 4 MILLION complaints over the last 5 years and a total loss of $37.4 billion [29]. Those are some huge numbers. Ok, so what happens if we dig into this data a little bit more? Perhaps break these down into sub-categories? In Table 2.2 you can see just SOME of the threats that we have talked about.

Essentially, this table tells something very important that we need to keep in mind: the number of cases does not equate to the financial amount of loss. For instance, while phishing had the most cases, it had the lowest losses, most likely meaning that the sundry of phishing attacks out there work, but only to a limited extent. On the other hand, investment fraud, which we talked about earlier, has only 39,000 cases but leads in the amount of money stolen. I'm going to finish up with one last statistic. And that is what exactly happens to these people once they've been compromised by a scammer? It's very clear to see what these

TABLE 2.2
US Fraud Statistics 2023 [29]

Scam	Number of Cases	Losses
Phishing	298,878	$18,728,550
Personal Data Breach	55,851	$744,219,879
Investment	39,570	$4,570,275,683
Tech Support	37,560	$924,512,658
Business Email	21,489	$2,946,830,270
Identity Theft	19,778	$126,203,809
Romance	17,823	$652,544,805
Advanced Fee	8,045	$134,516,577

TABLE 2.3
US Scam Cases and Losses in 2023 [30]

Outcome	Percentage Reported
Credit card is compromised	64%
Data is stolen	32%
Account is hacking	31%
Money is stolen	23%

scammers are after. You can see for yourself in Table 2.3. So, the takeaway is, if anyone asks for your credit card number, tread veeeery, veeeery carefully.

Now all this could all have been avoided if people were just a little more careful or more educated with the risks of online interactions, especially when giving away information or money. Let Mitnick's words continue to ring in your mind: the weakest part of any system will always be the human. We can't solve the human problem, but we can improve it. We can try to minimize the number of people that fall victim to a scam by learning about what scams might look like, being more wary about who we give our personal details and money to, and maybe even by putting effort into becoming more technologically fluent. This applies to both individuals and employees within a company (need I remind you about what happened with Snapchat?). Given all of the information in this chapter, I'm actually surprised that Monk hasn't fallen victim to an online scam. Monk is one of my favorite shows, but I always found it curious that online scams weren't shown. Given that he's technologically challenged, he would have been easy prey, especially because the whole show is riddled with instances where people abuse his condition. Perhaps this can just be attributed to the lack of general awareness at the time, given that the show was filmed between 2002 and 2009. But as he so eloquently put to Natalie in the episode entitled "Mr. Monk and the UFO" (as he was explaining what had happened to the murder victim) "the Internet People… they believe anything".

REFERENCES

[1] "Delphic oracle," Britannica, Jun. 7, 2024, [Online]. Available: https://www.britannica.com/topic/Delphic-oracle. [Accessed: Jun. 25, 2024].

[2] "The weakest link in safety is still man. Kevin Mitnick showed us how to outsmart us," Mitnick Security, Oct. 19, 2018, [Online]. Available: https://www.mitnicksecurity.com/in-the-news/the-weakest-link-in-safety-is-still-man.-kevin-mitnick-showed-us-how-to-outsmart-us. [Accessed: Jun. 25, 2024].

[3] "What is 'Social Engineering'?," ENISA, [Online]. Available: https://www.enisa.europa.eu/topics/incident-response/glossary/what-is-social-engineering. [Accessed: Jun. 25, 2024].

[4] "Kevin Mitnick: The World's Most Famous Hacker," Mitnick Security, [Online]. Available: https://www.mitnicksecurity.com/about-kevin-mitnick-mitnick-security. [Accessed: Jun. 25, 2024].

[5] "History of Phishing," phishing.org, [Online]. Available: https://www.phishing.org/history-of-phishing. [Accessed: Jun. 25, 2024].

[6] "What is phishing?," Malwarebytes Labs, [Online]. Available: https://www.malwarebytes.com/phishing. [Accessed: Jun. 25, 2024].

[7] A. Hern, "Snapchat leaks employee pay data after CEO email scam," The Guardian, Feb. 29, 2016, [Online]. Available: https://www.theguardian.com/technology/2016/feb/29/snapchat-leaks-employee-data-ceo-scam-email. [Accessed: Jun. 25, 2024].

[8] "Have you heard of the Whale Phishing Attack?," Neumetric, [Online]. Available: https://www.neumetric.com/have-you-heard-of-the-whale-phishing-attack/#:~:text=In%202016%2C%20Snapchat's%20payroll,payroll%20data%20to%20a%20scammer. [Accessed: Jun. 25, 2024].

[9] "19 types of phishing attacks," Fortinet, [Online]. Available: https://www.fortinet.com/resources/cyberglossary/types-of-phishing-attacks. [Accessed: Jun. 25, 2024].

[10] "Holy Water: a creative water-holing attack discovered in the wild," Kaspersky, Mar. 31, 2020, [Online]. Available: https://www.kaspersky.com/about/press-releases/2020_holy-water-a-creative-water-holing-attack-discovered-in-the-wild. [Accessed: Jun. 25, 2024].

[11] "Amazon Spoofer Attempts Credential Phishing with Look-alike Domain," Abnormal Security, [Online]. Available: https://intelligence.abnormalsecurity.com/attack-library/amazon-spoofer-attempts-credential-phishing-with-look-alike-domain. [Accessed: Jun. 25, 2024].

[12] "Phishing," Encyclopedia by Kaspersky, [Online]. Available: https://encyclopedia.kaspersky.com/glossary/phishing/. [Accessed: Jun. 25, 2024].

[13] Phil, "MPs Bombarded by Spam as Brexit No Deal Nears," Infosecurity Magazine. [Online]. Available: https://www.infosecurity-magazine.com/news/mps-bombarded-spam-brexit-no-deal/. [Accessed: Jun. 25, 2024].

[14] "5 Smishing Attack Examples Everyone Should See," SecureWorld, Sep. 2, 2019, [Online]. Available: https://www.secureworld.io/industry-news/5-smishing-attack-examples-everyone-should-see. [Accessed: Jun. 25, 2024].

[15] S. Ellis, *This present darkness: A history of nigerian organized crime*, Oxford University Press, 2016. [Accessed: Jun. 25, 2024].

[16] "Top Online Scams and How to Avoid Internet Scams," Kaspersky, [Online]. Available: https://me-en.kaspersky.com/resource-center/threats/top-scams-how-to-avoid-becoming-a-victim. [Accessed: Jun. 25, 2024].

[17] "The Most Common Photos Used by Catfish Scammers: Fake Love Costs Real Money," Hackernoon, Sep. 2, 2023, [Online]. Available: https://hackernoon.com/the-most-common-photos-used-by-catfish-scammers-fake-love-costs-real-money. [Accessed: Jun. 25, 2024].

[18] "Pig butchering scam," Malwarebytes Labs, [Online]. Available: https://www.malwarebytes.com/cybersecurity/basics/pig-butchering-scam. [Accessed: Jun. 25, 2024].

[19] G. M. Sykes and D. Matza, "Techniques of neutralization: A theory of delinquency," *American Sociological Review*, Dec. 1957, [Online]. Available: https://doi.org/10.2307/2089195. [Accessed: Jun. 25, 2024].

[20] A. T. Ebot, "Advance fee fraud scammers' criminal expertise and deceptive strategies: A qualitative case study," *Information and Computer Security*, Oct. 2023, [Online]. Available: https://doi.org/10.1108/ICS-01-2022-0007. [Accessed: Jun. 25, 2024].

[21] M. Offei, F. K. Andoh-Baidoo, E. W. Ayaburi, and D. Asamoah, "How do individuals justify and rationalize their criminal behaviors in online romance fraud?," *Information Systems Frontier*, Apr. 2022, [Online]. Available: https://doi.org/10.1007/s10796-020-10051-2. [Accessed: Jun. 25, 2024].

[22] J. N. B. Barnor, R. Boateng, E. A. Kolog, and A. Afful-Dadzie, "Rationalizing Online Romance Fraud: In the Eyes of the Offender," *AMCIS 2020 Proceedings*, Aug. 2020, [Online] Available: https://aisel.aisnet.org/amcis2020/info_security_privacy/info_security_privacy/21/. [Accessed: Jun. 25, 2024].

[23] G. Norris, A. Brookes, and D. Dowell, "The psychology of Internet fraud victimisation: A systematic review," *Journal of Police and Criminal Psychology*, Jul. 2, 2019, [Online]. Available: https://doi.org/10.1007/s11896-019-09334-5. [Accessed: Jun. 25, 2024].

[24] K. Parti and F. Tahir, "'If we don't listen to them, we make them lose more than money:' Exploring reasons for underreporting and the needs of older scam victims," *Social Sciences*, Apr. 28, 2023, [Online]. Available: https://doi.org/10.3390/socsci12050264. [Accessed: Jun. 25, 2024].

[25] G. Norris and A. Brookes, "Personality, emotion and individual differences in response to online fraud," *Personality and Individual Differences*, Feb. 1, 2021, [Online]. Available: https://doi.org/10.1016/j.paid.2020.109847. [Accessed: Jun. 25, 2024].

[26] M. Butavicius, R. Taib, and S. J. Han, "Why people keep falling for phishing scams: The effects of time pressure and deception cues on the detection of phishing emails," *Computers & Security*, Dec. 2022, [Online]. Available: https://doi.org/10.1016/j.cose.2022.102937. [Accessed: Jun. 25, 2024].

[27] D. M. Sarno and J. Black, "Who gets caught in the web of lies?: Understanding susceptibility to phishing Emails, fake news headlines, and scam text messages," *Human Factors*, May 1, 2024, [Online]. Available: https://doi.org/10.1177/00187208231173263. [Accessed:Jun. 25, 2024].

[28] N. Ebner, D. Pehlivanoglu, "Are older adults more vulnerable to scams? What psychologists have learned about who's most susceptible, and when," University of Florida News, Jun. 11, 2024, [Online]. Available: https://news.ufl.edu/2024/06/older-adults-vulnerable-to-scams/#:~:text=Individual%20risk%20factors&text=For%20example%2C%20among%20people%20around,greater%20susceptibility%20to%20email%20phishing. [Accessed: Jun. 25, 2024].

[29] Federal Bureau of Investigation, "Internet Crime Report 2023." Internet Crime Complaint Center, Mar. 2024, [Online]. Available: https://www.ic3.gov/Media/PDF/AnnualReport/2023_IC3Report.pdf. [Accessed: Jun. 25, 2024].

[30] "Nearly 1 in 3 Americans report being a victim of online financial fraud." Ipsos, Dec. 2023, [Online]. Available: https://www.ipsos.com/en-us/nearly-1-3-americans-report-being-victim-online-financial-fraud-or-cybercrime. [Accessed: Jun. 25, 2024].

[31] N. Low and C. Lally, "Social and Psychological Implications of Fraud." UK Parliament, Apr. 2024, [Online]. Available: https://researchbriefings.files.parliament.uk/documents/POST-PN-0720/POST-PN-0720.pdf. [Accessed: Jun. 25, 2024].

3 Me, Myself, and a Stranger – Chronicles of an Identity Thief

So, I just finished watching the Ashley Madison documentary on Netflix, and its ripe in my brain. Therefore, I'm going to start this chapter by telling you that story. For those of you that may not know, Ashley Madison (funnily enough just named after the two most popular girl's names in the US at the time) is a dating site catering to married people. The whole premise of "Life's short. Have an affair" is not new, however, no one has taken advantage of it quite like Ashley Madison. Now, at the peak of popularity, there was a claimed 60 million users on the site, in 53 different countries [1]. We say claimed, because that's what the company says, and given the rest of what I'm about to tell you, we have a few reasons to doubt their truthfulness. People signed up for this site using their names, credit card information, location, etc, just like they do for any other online service. When you do this, you kinda expect these services to protect your information, or at the very least not share or publicize it. That's generally in the implicit contract. Well, as you can guess, people (apart from the users probably) were not very fond of Ashley Madison. They deemed it crass, immoral, perverse, and uncouth against God. Therefore, logic would dictate that a company with so many haters would take every effort to protect the identity of its users. Unfortunately, that was far from the deep dark truth of Ashley Madison.

 DOI: 10.1201/9781032679389-4

Not only were customer data not secured properly, but Ashley Madison was also even charging people an extra $20 to "securely delete" their profile to ensure no traces of their being on the site remained, should they choose to leave the service. Needless to say, that also was never done. Of course, as anyone with a 5th-grade education would have expected, the website got hacked. Funnily enough, the hackers, calling themselves "The Impact Team" even gave Ashley Madison the option to prevent any data disclosure, by simply shutting the website down. Long story short, the Ashley Madison crew chose money over customers (this isn't unique to just them, so don't be surprised when Fortune 500 companies choose revenue over your data). We still don't know who The Impact Team is or was, perhaps jilted lovers of adulterers on the site, or just ultra-conservative religious zealots, or perhaps just a Russian 15-year-old in his bedroom having a good time (spoiler alert: they were never caught), but they didn't want money. They just wanted the site shut down (which at the end of the day is pretty commendable, wouldn't you say?). And to think they even gave numerous warnings to Ashley Madison…

So, warnings were made. Essentially, shut Ashley Madison down, or the hackers will leak all of the data from the servers, disclosing the personal information of every one of the users. Ashley Madison continued operating and assured everyone that their data was secure. At the same time, they hired two Swedish cybersecurity experts to try and run risk mitigation proactively, fully knowing that the customer databases were NOT secure. After a few days of non-compliance from the Ashley Madison team, the hackers decided to show that they weren't just playing around and released the personal information of 2,500 of the (at the time) 30 million users. They then gave Ashley Madison 30 days to shut down, or everything was going to be made public. Of course, now, Ashley Madison's top guns knew The Impact Team was a real threat, and clearly had access to the servers. So, what do they choose to do? Absolutely nothing. Everything was just swept under the rug, until one fateful day in August 2015, when 60 gigabytes of data was uploaded as a BitTorrent download, with the link to it available on the dark web.

Immediately everyone started tearing through the data, trying to find if their significant other, neighbor, teacher, daughter, lawyer had a profile on the site. As you can guess, a lot of things came from this type of data breach, everything from broken marriages, to lost jobs, to even suicides. Companies started making the data even more easily accessible by creating search engines so that people could look up the names of outed individuals [2]. Others began extorting money from affected individuals by saying if they didn't pay, they will publicly humiliate them. Essentially, the internet turned into a mob of drama-hungry teenagers, trying to infiltrate the Ashley Madison users' private lives.

Further compounding the craziness of this whole situation was that the internal emails of the Ashley Madison CEO and team were released in addition to the user data. Emails back and forth from the CEO admitting that the secure delete that users were paying $20 for (and didn't actually do anything), netted them over a $1 million a year in revenue. Turns out he wasn't as nice of a guy people had thought (*cough, cough*). What a surprise. Get this, the payment scheme of Ashley Madison in a nutshell was that women didn't pay anything, but men paid for everything like sending and viewing messages. Now, of the 30 million users at that time, it was revealed that 5.5 million were registered female accounts. Not the greatest guy-to-girl ratio, but still. Further digging showed that of these 5.5 million female users, only 0.2%, or about 12,000 actually used their account. The rest were created and forgotten (also, it was found that a good chunk of these could be traced back to being created using Ashley Madison office IP addresses). Then, of these 12,000 active female users, only 9,700 ever responded to a message from a male [1]. Clearly that is not a sustainable business model! So, what did Ashley Madison do? Create fake profiles of course! And guess who's pictures they used? Those of real-life women. But fake profiles don't respond to men, and if Ashley

Madison was making money off of men messaging, then this wouldn't work very well. So, what next? Creating AI-based bots to message for the fake women. Yep, that's right. Bots messaging men, and every time a man opened, read, or responded to a message from a bot, Ashley Madison made money. To further add insult to injury, there were cases of women being approached by men, telling them that they had been messaging on Ashley Madison. Of course, these women had no idea that their physical likenesses were being inhabited by bots [3].

The world is an absolutely insane place, and at this point, you're probably better off if nothing surprises you anymore. So, why is this Ashley Madison case so significant? Well, technically, it is two-fold. One, we can see how much sensitive and private data people share with random companies over the internet. And secondly, we can see just how easy it is for someone, be it a company, a bot, a hacker, or another person, to simply take over or ruin your life.

We've all done this. We've all shared too much information about ourselves at some point. But who is in the wrong at the end of the day regarding this specific case? Well, of course, the company for failing to properly secure customer's data. Also, for selling shit that didn't actually exist. Ohhh, and for duping people to spend millions and millions of dollars messaging fake bots. But what about the people themselves? The users? Were they innocent? Well, technically, they knew what they were getting themselves into. I'm sure there were a few people who didn't care about the data breach and could care less about being publicly outed as an Ashley Madison user. However, if I were a betting man (which I'm not, because every time I've tried to gamble, I always ended up losing), I would bet heavily on the fact that 99% of the users did not want their identities to be revealed. They did not want their spouses, their friends, their family knowing about their deepest, darkest polygamous secrets. And that is how we ended up with such extreme cases as suicide. And finally, are the hackers innocent? Absolutely not, the blood of these suicide victims, not to mention the mental torture of millions of innocent people who were blindly caught up in the drama of others (such as the spouses, for instance) are on The Impact Team's hands. But at the end of the day, what happened? Millions of lives ruined. Hackers never caught. And Ashley Madison sued in a class action suit for a measly $11 million (not even 10% of their annual revenue).

So, dear reader, what did we learn from all this? It is that the user always ends up losing the most when it comes to private data on the internet. "Only you can prevent forest fires", so says Smokey the Bear. But in this case, he means that only you can prevent your data from being leaked or used against you. Now, the Ashley Madison case, while involving some aspects of identity theft (like the use of bots masquerading as real-life people), is still not a cut-and-dry traditional identity theft case. But it does share a lot of similarities (hence why I spent like four pages talking about it) and gets us thinking and talking about just how bad things can be. The main common ground is the idea of what we call the 'prudential citizen' [2]. Basically, the notion that individuals should be responsible for their own actions and take actions to reduce the likelihood of a negative outcome. By uploading a whole heap of your private life on Ashley Madison, you probably are not being very prudent, are you? Websites and apps SHOULD expect individuals to be responsible for their own actions, and theoretically people SHOULD take necessary precautions to protect their identity. But is this true in cyberlife? The idea of the prudential citizen places the responsibility for any data breach on the victims themselves, for not taking necessary actions to prevent the incident from occurring. But what about children? What about the elderly? Or, simply the technologically challenged? The prudential individual hinges on the assumption that people understand what it is that they are doing, and the potential consequences of their actions. And that is one BIIIIG assumption. So, let's take a stroll down Identity Theft Lane, and visit all of the different houses that you can enter, if you, yourself, fail to be a prudential citizen. Buckle up, because we have quite a few stops to make… Everything from money, to scams, crimes, all the way to psychological torture, we got it all.

FOR MONETARY GAIN

Money, it's a gas. Grab that cash with both hands and make a stash.

Alrighty, let's start with the simplest and most common form of identity theft, money. Isn't that why we do most of the things in life anyways? Quick, simple, easy cash grabs. I say easy, because in the next section we will turn it up a notch from plain monetary gain, to fraud, scams, cons, and crimes (in addition to the monetary gain of course).

Well technically, before we get into modern-day identity theft, let's set the clock back a bit. Say, London, 1848. You're jogging down a back alleyway, and suddenly… THWACK! You're laying on the ground, your fractured skull bleeding out all over the cobblestones. You were Sir Zaphod Beeblebrox, but now you're no one. You're dead as a doornail. Your money and pocket watch ripped from your body right before your last exhale, along with the clothes off your back. Someone else is now Sir Zaphod Beeblebrox, and no one's the wiser. There were no photo IDs back then… perhaps an oil painting, if you were rich enough to have one commissioned. But even those tended to be a little "optimistic" (I guess much like Instagram filters of today). So, it was their word against yours, the only real proof to identity. There were of course signatures, again easily forged, and physical signet rings for melting wax seals onto letters, also easily stolen. In fact, there were no identification systems in place until the turn of the century. But then again, identification wasn't really needed. Business was cash based; loans weren't generally given out without collateral. The closest we got to a trust-based credit system (similar to what we have today) was getting a written letter of recommendation from your most prominent buddy saying that you could be trusted.

Then in the early 20th century, photographic identification became a thing (however, it wasn't really fully utilized until the middle of the century). Over the years, we also were issued a slew of numbers... passports, driving licenses, social security numbers, tax identification numbers, bank accounts, credit cards, addresses, and so forth. A person's identity slowly became a series of random digits. These random digits allowed you to buy a house, take out a bank loan, get a credit card, and make every-day purchases. Criminals quickly caught onto how easy it was to take these numbers from people, simply by tricking them. The first technological form of identity theft was using a telephone. In the 1960s and 1970s, nearly 100% of identity theft was done over the phone [4]. In these scams, identity thieves would call up unsuspecting victims and inform them that they had won an all-expenses paid round-trip ticket to Bermuda, or perhaps a 1968 Plymouth Barracuda with that 426 Hemi. All they needed to do was verify the identity of the lucky winner! Just a quick check of social security number, name, address, and bank account infor-mation, and that shiny new car would be all yours! Of course, there was no vacation, and no shiny new car, and the thieves would now have all of your information to take out a bunch of loans in your name. We can't be too harsh on these victims, as information about identity theft and the importance of safe-guarding your information at this time was not very prevalent and publicized. People just had no fear of disclosing this infor-mation, as "what's the worst that could happen?" Today, of course, the amount of this type of phone-based identity theft has decreased significantly and is only about 7% of ID theft in total, but that's still not zero. You'd think human beings could catch on to a 60-year-old scam...

After phone identity theft came the age of dumpster diving. Until the modern era, credit card numbers were not hidden on receipts and statements as they are today. There was no "XXXX XXXX XXXX 2524", everything just had the full credit card number written out on it! This gave perps the excellent opportunity to sift through people's trash and collect receipts and bank letters. As privacy was not an issue, they would have access to tons of different data sources, ripe for the picking. Throughout the 1980's no one even considered the possibility that someone would want to dig through their dirty, stinky trash. So, no countermeasures were taken. The news and media quickly caught on to this trend of criminals stealing people's identity and information through their trash. Thus, the paper shredder industry went nuts. Before the 1980s paper shredders were exclusively a government, military, and banking thing. Now, everyone and their mom went out and bought one [5].

And that brings us up to date. Today, 62% of identity theft occurs over the internet [4]. It's clear that identity theft has significantly changed the popularity of the internet. The landscape has shifted from overt theft to more sophisticated practices, but the primary objective remains, to steal personal information for money. Whether this is done because of your carelessness when buying that new PS5 on some sketchy website at half-price, or because you sent money to a Nigerian prince who happens to have his assets frozen due to some big misunderstanding but will repay you tenfold as soon as he clears it up. Either way, you voluntarily gave information to someone who shouldn't have it, and that's where it all starts (Table 3.1).

In the table, you can see the FCC's most common types of identity theft in 2023, and keep in mind, these are just the ones reported… think of how many cases go unreported per year as well. So, keep an eye out for people asking for social security numbers (or if you're not as American as apple pie, whatever equivalent your country may have), as that's how credit cards are opened, which is the most popular form of ID theft. Also, when online shopping or responding to emails (the second largest chunk of the table at #2 under miscellaneous identity theft), all I can say is USE YOUR BRAIN. Look at URLs (the web addresses themselves – sorry to say, but Amazon's URL is not Amaezon. cc, so you probably shouldn't enter your credit card number

TABLE 3.1
FCC's Top 5 Most Common Types of Identity Theft [6]

Type of Identity Theft	Number of Reports	Percent of Total Top Five
Credit card fraud-new accounts	381,122	42.0%
Miscellaneous identity theft*	279,221	30.7%
Bank fraud-new accounts	84,335	9.3%
Government benefits fraud-applied for/received	82,419	9.1%
Loan fraud-business/personal loan	81,342	9.0%
Total, top five	**908,439**	**100.0%**

* Includes online shopping and payment account fraud, email and social media fraud, and medical services, insurance and securities account fraud, and other identity theft.

there). And the same goes for email addresses, Jeff Bezos is NOT emailing you from a random Yahoo account asking to verify your social security number.

FOR SCAMMING OR COMMITTING A CRIME

Money, it's a crime. Share it fairly but don't take a slice of my pie.

Scams and crimes. Now, these two topics go hand-in-hand. Yes, we already have a whole other chapter on this topic, but I am going to avoid overlap and focus on just those involving identity theft. Simply put, scamming others (or committing a crime), all while assuming someone else's name and identity. There are a million reasons why someone would want to do these things, but in this section, we will focus on something a tad more sinister than the easy-to-pick apples in the first section. Here, we'll talk about everything from social engineering, to fraud, and even framing.

We've all heard of people who were tricked into doing something that they should not have done. This type of psychological manipulation is called social engineering. Essentially, by pretending to be someone in a position of power, you can easily trick a lower-level employee into revealing confidential information, or even giving you system or physical access. As the saying goes, "act like you own the place". This type of confidence trick often scares and confuses people, instead of asking for permission, or thinking about what they're doing, they just blindly give in and accept their fate (because if they're wrong, they end up looking stupid or reprimanded anyways, and could lose their jobs). There was a video a while back of Lewis Hamilton in Melbourne at the 2014 Australian Grand Prix that got me thinking about this exact type of scenario. So, Lewis had to retire from the race and rode a scooter back to the paddock, and of course had to pass through the security swipe gates for re-entry. He's fully kitted out, helmet, visor, full race gear. He tries to open the swipe gates, but clearly doesn't have his pass on him (I mean, who would if you had just gotten out of your F1 car). Security stops him, asks for his badge, and Lewis responds with "I'm a driver" and walks on in (if you Google "Lewis Hamilton I'm a driver" you'll find the video). Anyone fancy free F1 passes? All you need is a helmet and some race gear!

While Lewis's story is all fun and well, and no one got hurt, there are tons of other cases where in a similar manner people were scammed out of their life savings. Let's talk about the big two that everyone knows: Simon Leviev and Anna Sorokin. Well, looks like this is going to be a Netflix-heavy chapter (as if it wasn't already so). If you're not all caught up on your Netflix queue, I apologize in advance, you may just have some more stuff to add besides the Ashley Madison documentary.

Simon Leviev, otherwise known as The Tinder Swindler, and Anna Sorokin, of Inventing Anna, would make quite the fabulous duo. Simon built his entire persona around pretending to be the son and sole heir of Russian-Israeli diamond mogul Lev Leviev (who is very much a real person). Anna, on the other hand, ran under the name "Anna Delvey", also claiming to be a wealthy heiress. Two peas in a pod. While Anna "only" defrauded people of $275,000, Simon operated on a much larger scale (but we will get back to him later). Anna essentially took her fake heiress status, combining it with her "don't you know who I am" attitude and was able to social engineer her way through the upper-crust New York social scenes [7]. She fabricated financial documents to trick people into thinking she had a massive trust fund scattered around Germany and Switzerland... In addition, she also falsified a bunch of wire transfer confirmations to further cement this. Of course, this opened up opportunities a plenty. Using all of this information, along with some fake documents and counterfeit checks, she was able to deceive her friends and acquaintances into paying for her expenses (just asking to borrow money, but never paying them back), as well as even tricking banks and real estate agents into giving her money outright, or approving loans without collateral. Then, she used all of this money to live a ridiculously extravagant lifestyle, buying the fanciest couture, and staying in the world's most luxurious hotels.

At the end, people started noting her hostile nature (a bit too much of the confidence trick we talked about before, perhaps), as well as the fact that she only paid with cash and lived in hotels (and no one does this but criminals trying to hide their paper trails). Once people started catching on to her antics, she even had the audacity to retaliate... For instance, when Anna was asked to pay in advance for her hotel booking at 11 Howard (at which she exclaimed "no one else must do that"), she vengefully purchased the domain names corresponding to the hotel's managers names and tried to blackmail them for $1 million each to get them back [8]. I bet she was fun at parties. Well, at the end, she ended up being sent to prison (surprise, surprise), where she served three years before being deported to Germany. Now, she must serve the remainder of her sentence under around-the-clock house arrest, with the worst punishment ever: no social media!

All this, but when compared to Simon Leviev, Anna Sorokin, or Delvey, or whatever she calls herself, still seems like a saint. Compared to Anna's petty $275,000, Simon managed to clear a whopping $10 million through his insane cons and schemes [9]. From an early age, Simon was up to no good. He started small with check fraud, but the big stuff happened after he changed his name from Shimon Yehuda Hayut to Simon Leviev, to fool others into thinking that he was the heir to the billionaire Russian-Israeli businessman known as "The King of Diamonds" [10]. Now that his new identity was established, the Ponzi scheme can begin.

For those of you that need a little refresher, a Ponzi scheme is essentially just a trick where a person "invests" money to make a profit. However, the profit comes from the next victim's "investment", and thus the cycle continues. In the same way, Simon, pretending to be this ridiculously wealthy billionaire's son, seduced women around the world through whirlwind romances after meeting them on online dating sites like Tinder (I guess he wouldn't have been very successful on Ashley Madison). In the Netflix movie, the first victim, Cecilie, falls head-over-heels with Simon, but things get juicy when he convinces her that he is in grave danger and "[his] enemies are after [him]". She then proceeds to send tons of her own money to save him. Shortly after, two other women are introduced, Pernilla and Ayleen, both telling very similar stories. Simon hooks his victims with his charm and "wealth" (which once again is just an illusion from his fake identity), and starts up a new relationship. Each time, he leads them to believe he is this billionaire's son, and heir to this diamond empire. Once lulled into this false sense of security, Simon would request the women to wire him large sums of money, or let him use their credit card, because... you guessed it! He's in grave danger, and his enemies are after him.

So, what makes this a Ponzi scheme at the end of the day? That's the beauty of the whole ploy. Simon uses one victim's money to seduce the next. Fancy hotels, lavish gifts, Michelin star dinners, and private planes. Using Cecilie's money, Simon charms Pernilla. Then, using Pernilla's money he gets more money out of someone else. And so on, and so on. Once all of this finally caught up with him, he was sentenced to two years of prison in Sweden, and again another 15 months in Israel (however, he was let out early on good behavior after only 5 months, and with a fine to the tune of $40,000). That's a great deal after stealing $10 million, isn't it? Still would have almost the whole lot left after paying the fine! The actual real-life Leviev family also ended up filing a criminal suit against Simon as well, for libelous publications, infringing privacy, violating trademarks, and damaging the family's name. Naturally, nothing has come of this yet, and just goes to show how careful you have to be when trusting people. It's very rare that large punishments come to these Simon- and Anna-type individuals, in fact, quite the opposite. They get glamorized in documentaries, while their victims continue to struggle. So, all we can conclude here is, if the billionaire Leviev family isn't able to make a criminal charge stick against Simon who stole their family name and plastered it across the world, what odds do you think you have if someone swindles you?

FOR PSYCHOLOGICAL TORTURE AND FUN

Money, so they say. Is the root of all evil today.

Sometimes money just isn't enough. Sometimes individuals (most often with sordid pasts) want something else. To put it lightly, they are out for blood... they want to harm someone. Whether it's for revenge, jealousy, pure rage, or simply for fun, there have been lots of similar cases where individuals pretend to be someone else for their target's psychological destruction. From the lightest end of the spectrum, we could have a prank phone call pretending to be someone else... I remember this vividly from my high school days, back when we didn't have caller

ID, so we had to simply take people's words for who they said they were. We would have kids calling their friends pretending to be a girl that is into them, or perhaps even a parent trying to scare them. Either way, this would either instill something in the person being called (be it fear, love, lust, repulsion, etc.), but it would also start rumors about the individual supposedly making the call. "Did you hear Julie called Mark and said he was so cute, and that she wanted to play seven minutes in heaven with him at Danny's party on Saturday!" Clearly, Julie wouldn't be too happy about that kind of joke. Because I heard Mark looks like a troll.

Somewhere slightly beyond these simple pranks, we have some scams intended to harm people or businesses, ruin reputations, and defame public image. Instagram and Facebook, for instance, are a treasure trove of these types of scams. Anyone can open up a new account, and it doesn't have to be in their own name. I could open one up tomorrow in YOUR name and start messaging people as you. This actually happened to me, and I have to say, it was not fun. Apparently, our neighbor's nanny was in cahoots with this shady guy. Lucky me, I was fortunate enough to see some of this shady business going on outside our house, so I brought it up to my friend, the neighbor (her employer). We can call him George. George Costanza. Nonetheless, I was just looking out for George, who then promptly had a chat with the nanny to cut out any of this funny business in and or near his house. Well, the nanny went straight to the alleged boyfriend and told him that I had seen them, and that is why their fledgling love had to die. Naturally, he did not take too kindly to this and was dead set on getting this nanny fired in spite. He made it his life's mission for a solid six months to just mentally whittle us down, until the nanny would be fired, and she could join him again (in the unemployment line). So, he started by barraging us with messages on Facebook. As we all (my neighbor and his wife, as well as me and my wife) had both our Instagram and Facebook profiles on public, it did not take very long for him to find us. And find us he did.

Ok, no big deal, block and move on. But he didn't stop there. Oh nooo... He then learned how to make new accounts. He started by making duplicate accounts of his own name, and variations thereof, and continuing to send us hundreds of messages a day. Still not the end of the world, we blocked him on all of these new accounts and went on with our day. Once he realized that this wasn't working, he had to get a bit more creative. *Lightbulb*. He made accounts of the four of us. This guy went onto Instagram and Facebook, took our pictures, and created four new accounts with slight variations to the spelling of our names. Posted OUR pictures. And then the kicker... started messaging OUR friends. AS US! Luckily, his English wasn't that great. Actually, it was barely English at all, just a few consonants haphazardly typed after each other, so it was a pretty big red flag to most of our friends, who quickly sent us messages about these potential fraud accounts. We alerted Facebook, and guess what? We got a message saying these accounts did NOT go against the community guidelines, and therefore did not necessitate being disabled. So, this went on and on and on. We naturally at this point locked down all of our accounts and made them inaccessible (I highly suggest you too, dear reader, do this!). Until one day he just disappeared. Kind of anticlimactic, if you ask me. This whole ordeal lasted about six months, and as George had not fired the nanny, I think the guy just got tired, chalked it up as a waste of time, and moved onto some other poor sap.

We can take these "simple" scams, pranks, call them what you will, and turn them into full-blown schemes lasting decades. And that is what our next story is about. You may have heard of the Hollywood Con Queen, but in case you haven't, Apple TV+ has a great documentary miniseries about it (see, I'm not Netflix biased at all!). Imagine this – you're a young hopeful Hollywood actor (or in this case, could also be a photographer, chef, makeup artist, etc., but for now let's say actor), and you get a WhatsApp message from a famous director. He says that he has seen some of your work on another show that you did for HBO, and was very impressed. Sure, he says, you're not a big name yet, but he sees potential in you. He was so impressed in fact, he would like you to work with him on his next hit movie, which starts filming soon in Indonesia. All you have to do is a little character work, some physical training, and you're in. You will be the lead, he says. He believes in you. So, you guys keep chatting and even start Skyping. You bust your ass, you do his training plan, work on your acting skills, what have you, for half a year. Then it's off to Jakarta. Naturally, just like Hollywood operates, you have to pay for your own airfare and travels, but fear not, it will all be reimbursed. But it won't be. Soon you will find yourself halfway around the world, running random errands at the behest of this director, only to find out there is no movie. Never has been, never will be. You just wasted six months of your life working toward nothing.

This is how Hargobind Tahilramani operated. Through a variety of internet communication tools, he managed to impersonate dozens of Hollywood A-listers and scammed over 500 victims out of a grand total of $2 million over the course of a decade nearly (of course, the money was mostly on travel and other expenses, so not paid to Hargobind himself strangely) [11]. He could have easily scammed money out of these people. Many victims claimed they would have paid him, had he asked. But it was the lack of monetary expense on behalf of the victim that made them keep believing this was their big shot, this was the real deal. They say, had money been involved, the red flags would have gone up much sooner. Without this aspect of money, it didn't seem like a scam. After all, who scams someone for anything but money these days? It just seemed too elaborate for the minuscule payout. Hargobind did not do any of this for money. He did this for fun, he did it to feel like a big-shot Hollywood producer. He did all of this due to "his own failed attempts to become a person of significance in Hollywood" [11]. Now the ramifications of these types of scams were twofold. First, as we said above, the victims themselves are scammed out of money and time. But the secondary outcome is far more sinister. This was around the time of COVID-19, and there was another big case sweeping Hollywood. The "Me Too" movement, along with the arrest of Harvey Weinstein (a famous Hollywood film producer), who was accused by dozens of women for sexual misconduct, and eventually sentenced to 16 years in prison. Hargobind did impersonate these famous Hollywood celebrities, he also sexually assaulted some of his victims while under the assumed identity (over the phone and through messages). Naturally, the victims assumed these sexual advances were coming from the famous director, producer, screenwriter, etc. in question, and many reported these actions to news outlets, on their social media, and to the relevant authorities, thus inadvertently smashing the fraudster's assumed identity's reputation to pieces [12].

At the end of the day, it doesn't matter what you do. Someone can barge in and pretend to be you in any given situation. They can try to trick you by being someone else. What their goal is also varies… Some are after money, some are after your identity itself, and others may want to commit a crime in your name, or just cause you mental stress, annoyance, and anguish. To these ends, all you can do is be careful. Be careful who you share your personal information with. Be careful where you upload your private data. Be careful HOW MUCH of all of this stuff you are sharing. And of course, be careful who you trust. Do your research, don't be hasty. Don't fall for stupid tricks, or for a deal that seems too good to be true. In the end, it may help keep you, you.

REFERENCES

[1] "Ashley Madison data breach," Wikipedia. [Online]. Available: https://en.wikipedia.org/wiki/Ashley_Madison_data_breach. [Accessed: Jun. 19, 2024].

[2] C. Cross, M. Parker, and D. Sansom, "Media discourses surrounding 'non-ideal' victims: The case of the Ashley Madison data breach," *International Review of Victimology*, vol. 25, no. 1, pp. 53–69, 2019. [Online]. Available: https://doi.org/10.1177/0269758017752410

[3] B. Light, "The rise of speculative devices: Hooking up with the bots of Ashley Madison," *First Monday*, vol. 21, no. 6, Jun. 6, 2016. [Online]. Available: https://firstmonday.org/ojs/index.php/fm/article/view/6426/5525. [Accessed: Jun. 19, 2024].

[4] "History of identity theft," Identity Theft Scenarios. [Online]. Available: https://www.identity-theft-scenarios.com/identity-theft-facts/history/#:~:text=The%20First%20Identity%20Thefts,number%20and%20other%20personal%20information. [Accessed: Jun. 19, 2024].

[5] F. Knolton, "Shredding through time," SEM Shred. [Online]. Available: https://www.semshred.com/shredding-through-time/. [Accessed: Jun. 19, 2024].

[6] "Facts + Statistics: Identity theft and cybercrime," Insurance Information Institute. [Online]. Available: https://www.iii.org/fact-statistic/facts-statistics-identity-theft-and-cybercrime. [Accessed: Jun. 19, 2024].

[7] "Anna Sorokin," Wikipedia. [Online]. Available: https://en.wikipedia.org/wiki/Anna_Sorokin. [Accessed: Jun. 19, 2024].

[8] R. D. Williams, "'As an added bonus, she paid for everything': My bright-lights misadventure with a magician of Manhattan," Vanity Fair, Apr. 13, 2018. [Online]. Available: https://www.vanityfair.com/news/2018/04/my-misadventure-with-the-magician-of-manhattan. [Accessed: Jun. 19, 2024].

[9] L. Kranc, "The Tinder Swindler Simon Leviev pretended to be the king of diamonds?," Esquire, Mar. 1, 2022. [Online]. Available: https://www.esquire.com/entertainment/tv/a38955743/tinder-swindler-simon-leviev-true-story-where-is-he-now/. [Accessed:Jun. 19, 2024].

[10] M. Padin, "Inside Tinder Swindler Simon Leviev's stolen wealth – and new life as a business coach," Mirror, Feb. 3, 2022. [Online]. Available: https://www.mirror.co.uk/tv/tv-news/inside-tinder-swindler-simon-levievs-26127091. [Accessed: Jun. 19, 2024].

[11] O. B. Waxman, "Fake movie projects and phone sex: The story behind Apple TV+'s Hollywood Con Queen documentary," Time, May 8, 2024. [Online]. Available: https://time.com/6975176/hollywood-con-queen-apple-tv-true-story/. [Accessed: Jun. 19, 2024].

[12] R. Murphy, "Hollywood's mysterious "Con Queen" is now impersonating marvel executives," Slate, Jul. 15, 2019. [Online]. Available: https://slate.com/culture/2019/07/a-mysterious-con-artist-is-impersonating-female-marvel-executives-for-elaborate-sham.html. [Accessed: Jun. 19, 2024].

4 Lies, Likes, and LOLs – Social Media's Fake News Fiasco

> The media is fake… the media is really the word — I think one of the greatest of all terms I've come up with is fake. I guess other people have used it perhaps over the years, but I've never noticed it.
>
> –Former President Donald Trump in an interview with Mike Huckabee on Oct. 7, 2017

After reading the above by dear old Don, take a moment, look inward, and see how you are thinking. Are you thinking the former president of the United States is something of a narcissist for claiming he discovered fake news or are you thinking he is the oracle, the source of unbridled truth and the man who has given birth to the very concept of fake news? For now, take stock of your thoughts because it's important and points to something about why many are drawn to fake news in a way similar to how a fly is drawn to fresh fecal matter. Let's continue with Don for a moment. At the time of writing this, he has been convicted for his part in making hush payments to a former star of the adult movie industry. The trial played out in front of the world on traditional media outlets but also social media platforms. Don also has a number of other criminal charges to contend with. Without a doubt, he is going to be busy trying to fend these off. But Don is no stranger to being hauled under the spotlight.

DOI: 10.1201/9781032679389-5

His standard operating procedure is to dismiss any news from what he has called the 'lamestream media' that paints him in a negative light as fake news. If it makes him potentially look bad, he immediately labels it as fake. Furthermore, he does not seem to care about the source of what he calls fake news and, in fact, the more reputable the source, the more he seems to double down and attack it.

One major US newspaper, The Washington Post, conducted an interesting little study in response to the frequent presidential deriding of their own reporting about Don's somewhat troubled relationship with objective reality. They systematically and carefully tracked day by day the 'untruths' made by the former president during his tenure as POTUS. They can't come out and call them lies for what are likely to be bona fide legal reasons, but nonetheless, they reported that by the end of his presidency, Don had accumulated a rather whopping 30,573 'untruths' averaging 21 per day [1]. It's not just The Washington Post to avoid any partisan bias here. There is, in fact, a dedicated Wikipedia page entitled 'false or misleading statements by Donald Trump' and other mainstream media outlets like CNN have also provided tallies in respect of the magnitude of Trump's 'untruths' [2].

While people expect politicians to lie as a matter of course (to a degree at least), what makes Don different is not merely the scale of his 'untruthing' but that much of it has taken place through the medium of social media. When Twitter suspended his account from the platform in January 2021, he had close to 89 million followers, and over the 12 years Don was using Twitter, he tweeted approximately 57,000 times [3]. To my knowledge, to date, no-one has analyzed how many of these tweets contained 'untruths' but given what The Washington Post and CNN have previously reported, it is difficult to believe that there was no spillover. We do know however that in the 24 days following the date of the presidential election in 2020, Twitter added warning labels to 200 (about 30%) of President Trump's tweets or posts he retweeted indicating they contained false, disputed or misleading information. On Thanksgiving Day in 2020, Trump falsely tweeted, 'Just saw the vote tabulations. There is NO WAY Biden got 80,000,000 votes!!! This was a 100% RIGGED ELECTION.' He followed that up by claiming on Twitter that 'the 2020 Election was a total scam, we won by a lot (and will hopefully turn over the fraudulent result).' Given the 89 million followers that he has, that's a lot of millions of people receiving fake news right into their pockets. Because of this, Don is to fake news researchers what a petri dish is to a microbiologist: an interesting place to watch weird and strange stuff take shape and to see what it grows into. In the case of Trump's tweets, it grew into the capitol riots.

At a basic level, there are two things going on here: The first being that verifiable truths are considered fake, and the second is that obvious untruths are being peddled as true. How did we get to this point and what role has social media played? Why do millions of Americans not believe that climate change is real [4]? Is it possible that what Don tweeted in 2012 might have something to do with that? Back then he claimed that climate change was in fact created by the Chinese to hinder US competitiveness and in case you think that's fake news, check it out for yourself on Twitter. Or X. Or whatever Elon's calling it this week. It's the real Don by the way.

Claiming that global warming is a concept in this tweet is a bit like saying the Earth is flat and was made flat by aliens while we weren't looking. I'm presuming here that a large proportion of the almost 70,000 'likes' that Don got on that post equates to people saying, 'yep Don you're spot on and right…it's the Aliens at it again.' Arguments about a flat Earth aside though, there are only two possible explanations for this tweet. First, Don actually believes what he is saying and is spreading his beliefs around like fertilizer (misinformation) or second, he knows it to be false and is deliberately shoveling nonsense to the masses (disinformation). Both propositions are, as you hopefully will have realized, are a bit nerve wracking given that he once had the access codes for enough nukes to wipe out the planet, whether it is actually flat or not. The key questions that this tweet gives rise to are exactly the ones that captivate those who study the phenomenon of fake news; why do so many people believe fake news and how on earth could they believe Don about the Chinese inventing global warming? Do people believe it just because Don said it? Is it just confirming what they already believe? Does it have something to do with how social media works? But it's important to remember that fake news can be dangerous because it has real consequences, and those consequences are often not of the good kind.

Fake news of course isn't anything new and it's been around a hell of a long time. Except, according to the Fynn effect, we are actually all supposed to be getting smarter and, in theory at least, we should therefore be less susceptible to it [5]. Well, try that argument with the people who attended the Comet Ping Pong pizza restaurant in Washington DC on December 4, 2016. On that day, the restaurant was filled with families as it normally was each Sunday when a man walked into the restaurant and shot the place up. Luckily, no one was injured. However, the background and motive for his actions were tracked to completely fake news circulating on social media platforms. A month prior to the shooting, false tweets spread widely and quickly on the net claiming that this particular pizza restaurant was in fact the base for a pedophile sex ring involving the Democratic presidential candidate Hillary Clinton, a former Secretary of State, and members of her campaign team [6]. Posts with the hashtag '#pizzagate' began to appear rapidly and as the number of people who believed in the 'pizzagate' conspiracy grew and threats directed at the pizza shop increased, there were more and more confrontations with people who believed the fake news. Although social media outlets subsequently banned posts related to 'pizzagate,' the cat was out of the bag and the threats did not stop, culminating in the appearance of a 28-year-old North Carolina man (not you this time Florida), who showed up at the shop with a rifle to do his own 'investigation.' According to a New York Times interview with the suspect after his capture, he was a soft-spoken, polite man whose only intention was to rescue the children trapped in the shop. That's where fake news on social media can bring you to today. Pizza, guns, and kids; an American specialty.

WHAT MAKES SOCIAL MEDIA PERFECT FOR FAKE NEWS?

Most fake news creation and dissemination appears on social media platforms because it's a near-perfect breeding ground for a contagion of misinformation and disinformation for at least four reasons. The current occupiers of the Kremlin in Moscow realized this many years ago and have been using social media platforms extensively to spread fake news for quite some time. By the way, the German federal government has become so sick of this that they now have an official website dedicated to identifying and correcting Kremlin-sourced fake news being spread among German users of social media [7].

Firstly, almost anyone can with a bit of creative thinking generate and post fake news which can be made to look legitimate very easily and can reach thousands of users in a matter of minutes. The bad news is that you can expect a tsunami of fake news in the very near future. Creating fake news has become an awful lot easier given that AI platforms can now generate thousands of believable fake headlines in just a matter of seconds. These AI programs can even create the right visual images, which is always a must on social media and even flesh out the story for you.

Secondly, there are human factors at play. Fake news spreads faster than truthful/accurate content because people are up to 70% more likely to repost fake news content which means it reaches more people more quickly [8]. There is also an element of human vanity behind this and some basic psychology. A lot of fake news is perceived as novel and people who share novel information like to be seen as 'being in the know.' It's an unfortunate and depressing fact that some people are prone to embrace and disseminate wacky ideas because they believe it will bring them greater attention, esteem and status within their online social network. Another factor is that fake news contains content that is designed to elicit strong negative emotions such as anger, fear, and disgust. When these emotions get switched on, it becomes much harder to engage in critical reflective thought and we are more likely to respond reflexively and share that content [9].

Thirdly, the way in which social media sites are constructed plays a big part in the dissemination of fake news. You might be tempted to think that the people who spread fake news are biased or prone to believing fake news in the first place, which we will get to later, but it's not that simple. One reason is that heavy users of social media platforms simply habitually repost fake news but the critical bit here is understanding how the habit is formed in the first place. To understand the formation of the habit it's important to recognize that social media platforms do not incentivize reposting accurate information. They are not paragons of virtue though their community guidelines might lead you to think otherwise. Their goal is to keep users logged onto their accounts, to keep posting and sharing and those users who post and share frequently, especially sensational, eye-catching information, are the most likely to attract attention. The 'likes' and 'shares' then serve to reinforce that behavior. Unfortunately, that's the benchmark for many users. Furthermore, the algorithms used by social media platforms only prioritize engagement which is obviously their goal (i.e., likes, comments, shares, followers) as a signal of 'quality' and rank the most 'liked' content at the top of users' feeds. However, given that algorithms prioritize the popularity of information it results in overall lower quality of content on a platform [10]. A lovely example of the triumph of presentation over content.

A fourth reason is actually related to how social media companies themselves respond to fake news being posted on their platforms. It's important that you keep in mind that the primary goal of social media companies is to keep the user using the platform. Users generate revenue. It's that simple. If we go back to Trump's tweets about the 2020 election being stolen from him, you will remember that Twitter eventually permanently suspended his account (though for some reason Elon Musk rescinded this in 2022) but that was only after years of tweeting and reposting fake news about a litany of topics. These included that the Clintons had Jeffrey Epstein murdered to cover up their involvement in his sex crimes, that 81% of white people were murdered by black people and Donald Trump claiming that his telephones in Trump tower had been wiretapped by then-President Barack Obama [11]. And don't forget about global warming. None of these tweets got him suspended. But one is his most infamous tweets was about the death of a 28-year-old woman named Lori Klausutis in 2001 who worked in Scarborough's congressional office in Florida, USA. Unfortunately, she had died while in the office and though medical authorities stated her death was the result of a heart condition that caused her to collapse and hit her head on her desk, Trump tweeted the following in May of 2020: 'When will they open a Cold Case on the Psycho Joe Scarborough matter in Florida. Did he get away with murder? Some people think so....... Isn't it obvious? What's happening now? A total nut job!'

Remember, Trump was POTUS at the time and here he was explicitly suggesting that Joe Scarborough, an American television host and political commentator had committed murder. It was not a coincidence that Joe happened to be a rather big critic of Trump and his policies and was vocal about it. In fact, Don had been pushing this particular conspiracy theory for quite a while because he didn't like what Joe was saying about him on national TV. If you think it's bad that a sitting president can take to social media and accuse someone, with zero evidence, of murder, what happened next was far worse and demonstrates the reality of how social media companies can throw gasoline on the flames of fake news. Lori's husband, Timothy Klausutis wrote to the then CEO of Twitter asking for the company to remove tweets from Trump that pushed the conspiracy theory. Except, they didn't; even though, as Timothy pointed out, the tweets were a violation of Twitter's community rules and terms of service. They obviously were. Twitter refused to remove the posts however because they were doing what they wanted, which was to keep users stuck to the platform. The company has long maintained that prominent figures like Trump are to a large extent exempt from its content policies, because their tweets are of historical significance and often in the 'public interest.' [12]. However, in this case the notable lack of action by Twitter, who figuratively had been allowing Trump to get away with murder (and literally with accusations of murder) attracted a lot of attention and critical questions were being asked of the company. They responded by referring to making obtuse product enhancements and saying they were 'making efforts to label misleading information.' On the face of it, that sounds like a positive step except these efforts actually often have the result of making fake news spread faster and deeper. To understand why, focus on the word 'label' here.

HOW WARNING LABELS MAKE FAKE NEWS SPREAD FASTER

As an example, in November 2020, Twitter started to flag tweets by adding little blue labels at the end of tweets which reflected content that the presidential election was stolen from Trump. These labels included statements like 'this claim about the election is disputed' and 'this claim is potentially misleading' preceded by a circled exclamation mark. A bit like the warning on a packet of cigarettes. Be careful now. On November 12, 2020, Twitter announced that it had labeled roughly 300,000 election-related tweets using the terms 'disputed' and 'potentially misleading' but crucially it did not remove these tweets or block them from spreading. But let's not be negative Nancy just yet. The reality though was these now-flagged tweets actually spread further, for longer, and received more engagement than unflagged tweets [13]. In other words, the labeling of the tweets in the way they did, resulted in the net effect of turbocharging them. While it's possible that warning labels on social media posts may reduce people's willingness to believe false information, they don't appear to be effective in reducing its spread. Cue social media companies at this point throwing their collective hands in the air and claiming we did not know that would happen. We tried our best. Except you can be pretty sure they did because there has been plenty of research over the past few years showing the effect that warning labels help fake news grow legs. Twitter collects a whole lot of data on what happens on its platform and the company has a whole load of data scientists poring over multiple aspects of user behavior on the platform. If they really wanted to stop fake news, they would either label it straight up as fake or simply block it but that's not good for business. Social media companies are neither benign nor neutral: Don't let those funny cat videos fool you otherwise.

Mainstream media outlets might not be perfect but at least there are editorial processes, checks and balances and ultimately the legal system. The best you can hope for on social media is that a content moderator will identify fake news early and quickly or that an algorithm will pick up on it. But even if they do, it will most likely just get slapped with a warning and end up spreading faster anyway. The 'why' of fake news is itself a complicated question to answer and beyond the confines of what we are talking about here. However, throughout history, certain types of 'leaders' have shown a desire to undermine an independent media. These individuals certainly don't like public criticism and want to control the narrative about themselves and their deeds. It's not a coincidence that Don has been consistent about one thing: attacking the mainstream media under the artificial guise that it is the purveyor of fakeness. It certainly appears his agenda is to completely discredit the mainstream media because it's in his best interest to do so by shifting them to where he wants them to be. The near miracle that he has pulled off is to get millions of Republicans to go willingly along with him. In 2016, almost three-quarters of self-identified Republicans said they had at least some trust in national news organizations. That number dropped to one-third in 2021 and its probably even lower today [14]. Most have probably now subscribed to Don's own social media platform, Truth Social. Mission accomplished.

PATTERNS OF SOCIAL MEDIA USAGE FOR NEWS

When it comes to where Americans regularly get news on social media, Facebook currently outperforms all other social media sites. Three in ten US adults say they regularly get news there and around a quarter regularly get news on YouTube [15]. Half of all users of X (what was formerly known as Twitter) report that's the source of their news. Women are more likely to use social media platforms as their source of news which is unsurprising given that women in general spend more time on social media platforms to begin with. However, adults under the age of 30 are now almost as likely to trust information from social media sites as they are to trust information from national news outlets [16]. The downside of course is that this younger group is at greater risk of exposure to fake news than older generations because of their higher social media usage. It also seems to be the case that those who rely on social media to get news don't seem to do a very good job of detecting fake news. Some clever scientists at the University of Cambridge in the United Kingdom have devised an online test which assesses susceptibility to misinformation [17]. High scores mean you are good at detecting fake news and low scores obviously indicate the opposite. Overall, it seems that those who consume social media as their primary source of news do not do as well as those who used traditional sources of news. If we break it down by platform it looks like this. Half of those who get their news from Snapchat got the lowest scores, while users of Don's social media platform Truth Social were second worst with 45%, followed by WhatsApp with 44%, TikTok with 41% and Instagram with 38% [18].

You might be thinking at this point that social media and fake news is just a toxic relationship, but the really bad news is that a lot of people who rely on social media fail to distinguish fact-based news from fake news: Some depressing estimates are that it's as high as 70%. That should raise some eyebrows. It also raises a whole series of questions. Who falls for this stuff? What makes people susceptible to fake news? Is it a certain kind of person? Is it due to political allegiances? Is it proneness to belief in conspiracy theories? Is it even the result of where you live? Well, the best approach to this is to think of it like a recipe, a combination of things that can make you susceptible because it is rarely a single thing. Starting with where you live, intuitively you might think that countries with despots in charge would be top of the pile for traditional media distrust. In fairness, you could not be found at fault for coming to that conclusion. If I were Libyan, I would have found myself somewhat hesitant to believe the media when Muammar Gaddafi was running the show at the time. Yet the USA, France and the United Kingdom lead the way on the lowest rankings of mainstream media trust [19]. (Finland by the way is the place to be for most trust but the weather is a bit off-putting for people to move there). To answer why this situation has come about, it is worthwhile remembering something that happened in following Don's inauguration as president in January 2017.

Size matters to Don, (bigly as he would say), and because it matters, he rolled out his press secretary, a guy called Sean Spicer, who stated that the crowd attending the inauguration was the 'largest audience to ever witness an inauguration – period – both in person and around the globe' [20]. However, there was a small problem, and it was the fact that aerial photographs of the crowd attending the event and D.C Metro ridership figures for the day clearly demonstrated that the crowd at Barack Obama's inauguration in 2013 was substantially larger. Ok, so maybe the guy made a mistake you might think, but where things got super strange was the day after Sean made his wildly inaccurate claim, the US counselor to the president, Kellyanne Conway, appeared at a press conference and things took a truly Orwellian turn. She breathed life into Orwell's concept of '*doublethink*' by referring to 'alternative facts' to support Sean's claim. Kellyanne provides what is perhaps one of the clearest examples of trying to push the disease of what is known as 'post truth'; a move away from the pursuit of objective truth, a diminishing trust in science, and the abandonment of evidence-based reasoning. This is the perfect swamp-like condition for fake news to thrive and here was a key government advisor advocating that truth is effectively whatever you want it to be. Maybe there is an alternative to gravity if she fell off the capitol building. It's also no surprise that in the four weeks following her public gaslighting of the American people, sales of Orwell's dystopian novel 1984 increased by 95,000%. At least some people recognized the bullshit.

MEDIA DISTRUST AND FILTER BUBBLES

In countries where distrust in traditional news media is increasingly promoted as Don (but also others) has been doing for years, a kind of selective exposure to news takes place and people can turn to social media platforms that distribute disinformation combined with a decrease in the critical evaluation of these sources [21]. More simply put, if you live somewhere where trust in the official media is on its knees, then people are more likely to turn to online digital sources to seek an explanation for the events going on around them. Except the problem with social media and the news is that social media is a prime example of what is called 'narrow casting.' The broad in broadcasting is there to signify that at least, in principle, it is attempting to reach a broad segment of the population. As you might have guessed, that is not what happens on social media platforms. Trump posts to his followers which is already a narrow group (albeit a large following), and the algorithms produce a filter bubble where you only typically see more of the same. So much for the great hope of the democratization of information.

Getting back to the explanation seeking for events because that's a critical thing here. We as humans are hardwired to seek explanations for events such as 'why is the summer feeling hotter' to 'where did COVID-19 come from?' This 'explanation seeking' mechanism of ours goes on speed when perceptions of uncertainty and threat are thrown into the mix. During the early stages of the COVID-19 pandemic, many of us were freaked out with elevated stress and anxiety and it's no wonder that the internet lit up with fake news that the virus was deliberate biological warfare or that 5G network towers were the culprit. In order to reduce the stress and regain a sense of control, we try to make sense. However, the problem is that under conditions of threat, we can become more susceptible to fake news and the sense we create isn't very sensible. Human thinking seems to go badly awry. Where do you go when you want information about something happening like COVID-19? It's not your local library or neighborhood professor you turn to. It's the device in your pocket and the number one thing people do on the internet is look for information. But not all information is created equally.

THINKING STYLES, HEURISTICS AND SUSCEPTIBILITY TO FAKE NEWS

We all like to think that we are good at thinking. I've rarely met anyone who has said 'yeah thinking isn't my strong suit…best leave it to someone else.' And yet our own thinking is far from infallible. Is that the reason why so many people believed Don when he said global warming was just an idea fabricated by the Chinese to undermine the USA? Is there a problem in our thinking that relates to not being able to spot fake news? Well, here is a question for you to figure out the answer to but watch your own thinking for a moment. A bat and a ball cost $1.10 in total. The bat costs $1.00 more than the ball. How much does the ball cost? My immediate answer when I first came across this was to say the ball cost 10 cents and a bat cost a dollar. I'm guessing that maybe you did too. We intuitively think (and very quickly by the way if you paid attention to your thinking) that's the correct answer because it makes intuitive and immediate sense. Easy peasy lemon squeezy, except it's the wrong answer. The actual answer is 5 cents for the ball and a dollar and 5 cents for the bat. Now your brain is going 'what... that can't be right.' If the ball costs a dollar more than the ball and if the ball is 10 cents it means the bat costs 1 dollar and 10 cents, but you still must add the cost of the ball which now brings the total price up to $1.20. If you think more carefully and devote some attention to it, you work out that the ball must be 5 cents and the bat $1.05 and 5 cents which hey presto brings it up to $1.10. This little mathematical conundrum demonstrates that there is in fact two types of thinking that we rely on: a quick easy intuitive one and a slow deliberate analytical one that requires more energy and attention. But the message is that the quick intuitive one can be wrong even if it feels right. To see an example of this in practice, all we have to do is return to Don, the gift that just keeps giving.

In early October 2020, posts on social media began to appear quickly claiming that a new Centre for Disease Control (CDC) study had found that the majority of people infected with COVID-19 wore masks. On October 15, 2020, Don made the classic bat and the ball mistake when in a town hall meeting broadcast by NBC, he stated, 'but just the other day, they came out with a statement that 85% of the people that wear masks catch it.' By 'they' he of course meant the CDC. Don seems to have reached the conclusion rather quickly that wearing a mask made you catch COVID-19 or at the very least offered no protection at all. While Don did use the correct percentage from the study, based on intuitive thinking (some might say limited or none) he misinterpreted what was really going on. In fact, the

CDC study reported that 85% had reported wearing masks always or often. However, the study also found that for those in the group who had tested negative, 89% had reported wearing masks with the same frequency. The study went on to point out that 'people with and without COVID-19 had high levels of mask use in public. Even for those who always wear a mask, there are activities where masks can't be worn, like eating or drinking. People with COVID-19 were more likely to have eaten in a restaurant.' Of course, it's impossible to 'always' wear a mask because how else could someone eat or drink and in fact being in a restaurant taking your mask off was the critical bit here. The poor mask was falsely accused and innocent.

The types of human thinking that the bat and ball test demonstrate have some fairly big implications. Maybe all presidential candidates should be given it. This is not to say that intuitive thinking is necessarily a bad thing, or, for that matter, analytical thinking is superior but it's certainly the case that we should be aware of the differences and pitfalls of both for ourselves. Perhaps more importantly we should know the situations which is preferable for the situation at hand. One of my all-time favorite journal articles about how people think about what they read has the absolutely wonderful non-academic title of 'On the reception and detection of pseudo-profound bullshit' [22]. In a truly wonderful experiment, the authors defined pseudo-profound bullshit as stuff that on the surface seems impressive, true and meaningful but is actually vacuous (a polite way of describing bullshit). Sometimes academics are pure wordsmiths. Anyway, they did something simple and presented people with bullshit statements consisting of buzzwords randomly organized into statements with a solid syntactic structure, but which had no discernible or logical meaning (e.g., 'Wholeness quiets infinite phenomena'). For your pleasure, here are some other examples of AI-generated woo-hoo dribble: 'The dreamtime is approaching a tipping point' and 'Without non-locality, one cannot grow' and my personal favorite 'Humankind has nothing to lose. We are in the midst of a high-frequency invocation of empathy that will align us with the universe itself. We are at a crossroads of transcendence and ego.' [23]. They then came along and measured thinking styles, specifically whether people had a predominantly intuitive style or a predominantly analytical style. Now, if you are receptive to the kind of bullshit statements above, the news is not good I'm afraid. The actual evidence is that those who are receptive to bullshit primarily rely on this intuitive thinking style but are also less reflective, lower in intelligence, more prone to believing conspiratory theories, are more likely to hold religious and paranormal beliefs, and are more likely to endorse complementary and alternative medicine. The ideal dinner date for some perhaps. By the way, for those of you who may have forgotten, Don did suggest once on national TV that injecting disinfectant might be a cure for COVID-19 [24]. Not quite the alternative medicine that most would have in mind, but I'm pretty sure that someone somewhere followed his advice. I hope they survived to read this. As the idiom goes: 'It pays to keep an open mind, but not so open that your brains fall out.'

An intuitive thinking style is a kind of mental shortcut people use when evaluating information but what causes it to actually kick in? It tends to switch on when we come across a piece of information where signs of authenticity or fakeness are not immediately available and when we have information overload like when scrolling through a feed. This is where something called heuristics comes into play which is a fancy term for mental shortcuts much like we encountered with the cost of the bat and the baseball. None of us are a blank slate and we all have prior beliefs and ideological predispositions. Well one of the ways these shortcuts come about is simply through exposure or how many times someone has been exposed to a particular piece of information. Repeated exposure makes something familiar and unbelievably even a single prior exposure to a fake news headline can actually increase later belief in the headline, even if it's extremely implausible [25]. Given that people typically have used multiple social media platforms (on average almost seven [26]), it's easy to see how repeated exposure to the same piece of news can result in feelings of familiarity and bring about what is known as the 'illusory truth' effect. Add to this, the fact that fake news typically spreads on social media platforms up to seven times faster than accurate news, familiarity can become established very quickly and easily.

HOW ALGORITHMS FEED FAKE NEWS

The problem though is not all human when it comes to belief in fake news or denialism or factual news. Everyone should remember that social media does not exist for your benefit. In reality, it's an income stream for its owners and they make their money by inserting paid-for advertising in between the images and videos that you like. The way in which social media platforms are designed enhances familiarity and often perpetuates an 'echo-chamber' effect. As mentioned earlier, these platforms are designed to keep you scrolling by giving you more and more of what you like. You are reinforced to stay on the feed because they use algorithms that look for, learn from, and implement patterns of your likes. They know that feeds that are emotionally provocative (both positive and negative) hold you longer and they are particularly good at identifying content that is similar to what has captured your attention previously. Once they learn, and they learn quickly, they serve up more of the same with some tweaks just to keep it a little fresh. Now what you have is a circular feedback loop that traps you in your own filter bubble [27]. Without even possibly realizing it, you're now in an echo chamber where you only get exposed to information or opinions that reflect and reinforce your own. There is proof of this in case you are wondering. In 2015, researchers from Facebook published a study which demonstrated that Facebook algorithms do in fact create an echo chamber among users by sometimes hiding content from individual feeds that users possibly would disagree with: for example, the algorithm removed one in every 13 diverse content feeds from news sources for self-identified liberals. Overall, the results indicated that the Facebook algorithm ranking system caused approximately 15% less diverse material in users' content feeds and a 70% reduction in the click-through-rate of the diverse material [28].

Another little trap we fall into without noticing is what's called the 'bandwagon effect.' We are often more heavily influenced by others than we realize. For example, the more 'likes' a particular piece of fake news accumulates on social media platforms, people are more likely to believe it simply because we are susceptible to social consensus- if a lot of others think it's true then it must be true. Basically, if we see that lots of other people think something is true, then we are more likely to think it's true too. Finally, and back to Don again, if you view the source of fake news as credible in the first place, then you are again more likely to believe fake news like global warming is a Chinese ruse to undermine the US economy. One big predictor of belief in fake news and denialism of accurate news is simply if the news content aligns with people's prior beliefs and ideological predispositions. Social media exacerbates this because as we just saw, it will happily deliver up more servings of your own particular brand of belief and ideology.

PERSONALITY TYPES AND SUSCEPTIBILITY

This intuitive type of thinking seems to predispose people to believing fake news but here is a final question: Are thinking styles associated with certain personality types? And what personalities are particularly prone to the consumption of, and belief in, fake news when busily scrolling away? Well, several studies suggest that individuals with certain pathological personalities (i.e., schizotypal, paranoid and histrionic types) are ineffective at detecting fake news but are also more vulnerable to experiencing more anxiety as a result of fake news [29]. Individuals with these disorders tend to have a lot of what's called 'magical ideation' which is a tendency to hold beliefs that unrelated events are causally connected despite the absence of any plausible causal link between them. Of course, it's not hard to see why paranoia is linked to belief in fake news. Endorsement of fake news that reflects conspiracy theories is a common feature of paranoid personalities in the first place and science-based information is often rejected because it is seen as part of the 'system' and controlled by the political system. Intuitive and analytical thinking styles are not fleeting. They tend to be stable across time and, interestingly, are linked to certain personality traits. For example, intuitive thinking styles are typically associated with extraversion and a motivation to experience new things, change, and excitement whereas individuals with a more analytical thinking style are motivated to find consistent, predictable patterns in their lives [30]. The intuitive types might be the ones to spend a night on the town with though.

SO, WHAT HAVE WE LEARNED?

Fake news is never going to go away but the reality is that social media is now the primary place where it plays out. Based on the combination of human frailties and how they interact with characteristics and intentional design characteristics of social media platforms, social media is now the perfect trojan horse for fake news to get into the heads of potentially millions of people. Given the potential for AI to drown the world in fake news and social media platforms to deliver it straight into your hand, there is an argument that there is a rather pressing need to do something about it. It's attempted manipulation on a scale never seen before but unfortunately it is extremely difficult to convince people they should exercise critical thinking and develop digital literacy in an online near-constant feed of cute videos of cats, celebrity gossip and seeing whatever gives you that dopamine rush to the nucleus accumbens when you're scrolling. Some

argue that it's up to social media platforms to combat misinformation and fact-check what some researchers have called
individual differences on 'knowledge vulnerability,' which just
seems like a polite way of saying help protect the people who
don't know their ass from their elbow [31]. It might be time to
learn the difference because social media companies do not have
that as their first priority.

REFERENCES

[1] G. Kessler, S. Rizzo and M. Kelly. "Trump's false or misleading
     claims total 30,573 over four years," The Washington Post, Jan.
     24, 2021. [Online]. Available: https://www.washingtonpost.com/
     politics/2021/01/24/trumps-false-or-misleading-claims-total-
     30573-over-four-years/. [Accessed: Jun. 22, 2024].

[2] "False or misleading statements by Donald Trump." Wikipedia,
     [Online].   Available: https://en.wikipedia.org/wiki/False_or_
     misleading_statements_by_Donald_Trump.   [Accessed: Jun.
     22, 2024].

[3] A. Madhani and J. Colvin, "A farewell to @realDonaldTrump,
     gone after 57,000 tweets," Associated Press, Jan. 9, 2021.
     [Online]. Available: https://apnews.com/article/technology-joe-
     biden-donald-trump-social-media-aba9e7c118b21910817f26d
     0a34a4e27. [Accessed: Jun. 22, 2024].

[4] A. Uteuova. "Nearly 15% of Americans don't believe climate
     change is real, study finds," The Guardian, Feb. 14, 2024.
     [Online]. Available: https://www.theguardian.com/us-news/2024/
     feb/14/americans-believe-climate-change-study. [Accessed: Jun.
     22, 2024].

[5] E. Myers. "The Flynn Effect – Explaining Increasing IQ Scores,"
     Simply Psychology, Aug. 7. 2023. [Online]. Available: https://
     www.simplypsychology.org/flynn-effect.html. [Accessed: Jun.
     22, 2024].

[6] "Real Events Casued by Fake News in the US." Marubeni
     Research Institute. [Online]. Available: https://www.marubeni.
     com/en/research/potomac/backnumber/19.html.   [Accessed:
     Jun. 22, 2024].

[7] "Examples of Russian disinformation and the facts." Bundesministerium des Innern und für Heimat, [Online]. Available: https://www.bmi.bund.de/SharedDocs/schwerpunkte/EN/disinformation/examples-of-russian-disinformation-and-the-facts.html. [Accessed: Jun. 22, 2024].

[8] S. Vosoughi, D. Roy, and S. Aral, "The spread of true and false news online," *Science*, vol. 359, no. 6380, pp. 1146–1151, 2018. [Online]. Available: https://doi.org/10.1126/science.aap9559. [Accessed: Jun. 22, 2024].

[9] "How it spreads." University of Victoria. [Online]. Available: https://libguides.uvic.ca/fakenews/how-it-spreads. [Accessed: Jun. 22, 2024].

[10] G. Ceylan, I. A. Anderson, and W. Wood, "Sharing of misinformation is habitual, not just lazy or biased," *Proceedings of the National Academy of Sciences of the United States of America*, vol. 120, no. 4, p. e2216614120, 2023. [Online]. Available: https://doi.org/10.1073/pnas.2216614120. [Accessed: Jun. 22, 2024].

[11] S. Min Kim and H. Knowles. "Trump retweets conspiracy post tying Clintons to Epstein's death," The Washington Post, Aug, 11, 2019. [Online]. Available: https://www.washingtonpost.com/politics/trump-retweets-conspiracy-post-tying-clintons-to-epsteins-death/2019/08/10/e9461004-bbbf-11e9-aeb2-a101a1fb27a7_story.html. [Accessed: Jun. 22, 2024].

[12] T. Spangler. "Twitter Explains What It Would Take for Trump's Account or Tweets to Be Deleted". Oct. 16, 2019. [Online]. Available: https://variety.com/2019/digital/news/twitter-donald-trump-account-deleted-policy-1203372720/ [Accessed: Jun. 22, 2024].

[13] Z. Sanderson, M. A. Brown, R. Bonneau, J. Nagler, and J. T. Tucker, "Twitter flagged Donald Trump's tweets with election misinformation: They continued to spread both on and off the platform," Harvard Kennedy School (HKS) Misinformation Review, 2021. [Online]. Available: https://doi.org/10.37016/mr-2020-77. [Accessed: Jun. 22, 2024].

[14] C. Cilliza. " Here's Donald Trump's most lasting, damaging legacy," CNN, Aug. 30, 2021. [Online]. Available: https://edition.cnn.com/2021/08/30/politics/trump-legacy-fake-news/index.html. [Accessed: Jun. 22, 2024].

[15] "Trust in America: Do Americans trust the news media?" Pew Research Center, Jan. 5, 2022. [Online]. Available: https://www.pewresearch.org/journalism/2022/01/05/trust-in-america-do-americans-trust-the-news-media/. [Accessed: Jun. 22, 2024].

[16] J. Liedke and J. Gottfried. "U.S. adults under 30 now trust information from social media almost as much as from national news outlets," Pew Research Center, Oct. 27, 2022. [Online]. Available: https://www.pewresearch.org/short-reads/2022/10/27/u-s-adults-under-30-now-trust-information-from-social-media-almost-as-much-as-from-national-news-outlets/. [Accessed: Jun. 22, 2024].

[17] "YourMist Streamlit app." YourMist. [Online]. Available: https://yourmist.streamlit.app/. [Accessed: Jun. 22, 2024].

[18] C. Murray, "Gen Z and Millennials more likely to fall for fake news than older people, test finds," Forbes, June 28, 2023. [Online]. Available: https://www.forbes.com/sites/conormurray/2023/06/28/gen-z-and-millennials-more-likely-to-fall-for-fake-news-than-older-people-test-finds/. [Accessed: Jun. 22, 2024].

[19] A. Fleck. "Where trust in the news is highest and lowest," The Wire, Jun. 3, 2023. [Online]. Available: https://thewire.in/media/where-trust-in-the-news-is-highest-and-lowest. [Accessed: Jun. 22, 2024].

[20] N. Rattner. Trump's election lies were among his most popular tweets, CNBC, Jan. 13, 2021. [Online]. Available: https://www.cnbc.com/2021/01/13/trump-tweets-legacy-of-lies-misinformation-distrust.html. [Accessed: Jun. 22, 2024].

[21] E. Aïmeur, S. Amri, and G. Brassard (2023). Fake news, disinformation and misinformation in social media: a review. *Social Network Analysis and Mining*, vol. 13, no. 1, p. 30. https://doi.org/10.1007/s13278-023-01028-5

[22] Pennycook, G., Cheyne, J. A., Barr, N., Koehler, D. J., and Fugelsang, J. A. (2015). On the reception and detection of pseudo-profound bullshit. *Judgment and Decision Making*, vol. 10, no. 6, pp. 549–563. https://doi.org/10.1017/S1930297500006999

[23] S. Pearce, "Bullshit Generator." [Online]. Available: https://sebpearce.com/bullshit/. [Accessed: Jun. 22, 2024].

[24] "Coronavirus: Outcry after Trump suggests injecting disinfectant as treatment". Apr. 24, 2020. [Online]. Available: https://www.bbc.com/news/world-us-canada-52407177 [Accessed: Jun. 22, 2024].

[25] T. J. Smelter and D. P. Calvillo, "Pictures and repeated expo-
     sure increase perceived accuracy of news headlines," *Applied
     Cognitive Psychology*, vol. 34, pp. 1061–1071, 2020. https://
     doi.org/10.1002/acp.3684

[26] B. Dean. "Social media users." Backlinko, Feb. 21, 2024.
     [Online]. Available: https://backlinko.com/social-media-users.
     [Accessed: Jun. 22, 2024].

[27] J. Fournier. "How algorithms are amplifying misinformation and
     driving a wedge," The Hill, Nov. 10, 2021. [Online]. Available:
     https://thehill.com/changing-america/opinion/581002-how-
     algorithms-are-amplifying-misinformation-and-driving-a-
     wedge/. [Accessed: Jun. 22, 2024].

[28] Z. Tufekci. "How Facebook's algorithm suppresses content
     diversity modestly—how the News Feed rules the clicks,"
     Medium, May 7, 2015. [Online]. Available: https://medium.
     com/message/how-facebook-s-algorithm-suppresses-content-
     diversity-modestly-how-the-newsfeed-rules-the-clicks-
     b5f8a4bb7bab. [Accessed: Jun. 22, 2024].

[29] Á. Escolà-Gascón, N. Dagnall, A. Denovan, K. Drinkwater, and
     M. Diez-Bosch, "Who falls for fake news? Psychological and
     clinical profiling evidence of fake news consumers," *Personality
     and Individual Differences*, vol. 200, p. 111893, 2023. [Online].
     Available: https://doi.org/10.1016/j.paid.2022.111893. [Accessed:
     Jun. 22, 2024].

[30] L. Sagiv, A. Amit, D. Ein-Gar, and S. Arieli, "Not all great minds
     think alike: systematic and intuitive cognitive styles," *Journal
     of Personality*, vol. 82, no. 5, pp. 402–417, 2014. [Online].
     Available: https://doi.org/10.1111/jopy.12071. [Accessed: Jun.
     22, 2024].

[31] "The dark psychology of social networks." PsychVarsity, May
     27, 2023. [Online]. Available: https://www.psychvarsity.com/
     The-Dark-Psychology-Of-Social-Networks. [Accessed: Jun.
     22, 2024].

5 Hacking Health – The Internet's Influence on Medical Information

Once upon a time, long, long ago (let's say December 1998), in a land far far away (let's say Alabama), people used to go to the doctor's office when they felt sick. Simple enough, right? One such person could be Joe-Bob. Joe-Bob went to see Dr. Wilkinson, a registered General Practitioner in the great god-fearing state of Alabama. Cowboy boots and Stetsons and all, yee-haw. So, Dr. Wilkinson, after taking a long, hard look at Joe-Bob and running some tests, says that Joe-Bob has lung cancer. Probably from the three decades of Cuban cigars Joe-Bob smoked when he was stationed in Guam during Dubya Dubya Two. If my math is correct, that would make Joe-Bob nearly 80 years old at this point, but still, just bear with me here. Joe-Bob has to undergo some chemotherapy, which is a painfully nasty process, in addition to the slew of other pain he is feeling from the cancerous cells swimming through his body. So, the good Dr. Wilkinson, in order to ease Joe-Bob's pain, prescribes a nice shiny new medication called OxyContin, something that would solve all of Joe-Bob's problems.

"That's great!" Joe-Bob exclaimed, after being prescribed his pills. He had just sat for half an hour in Dr. Wilkinson's office and actually learned quite a bit about OxyContin through a

 DOI: 10.1201/9781032679389-6

promotional video running non-stop in the waiting room. "They don't wear out; they go on working; they do not have serious medical side effects," the physician in the video said, "these drugs [...] are our best, strongest pain medications" and "should be used much more than they are for patients in pain" [1]. On top of that, Purdue Pharma, the company that created OxyContin also said that a single dose relieves pain for 12 hours, more than twice as long as other types of pain medications. So, Joe-Bob would not have to wake up in the middle of the night to take his regular pain pills like usual, as one OxyContin tablet in the morning and one before bed would provide "smooth and sustained pain control all day and all night" [2].

Well, after going home and trying out his new-fangled medicine, Joe-Bob quickly realized it wasn't all that it was cracked up to be. He found out that the pills wore off hours early, leaving Joe-Bob writhing in agony for much of the day. Well, that is because essentially, OxyContin, for all intents and purposes, is a chemical cousin of heroin. And when it doesn't last, people like Joe-Bob go into excruciating bouts of withdrawal, including a higher craving for more of the pills. So, Joe-Bob goes back to Dr. Wilkinson, who then prescribes a higher dose of OxyContin, and the vicious cycle continues, until the worst happens. Luckily, our friend Joe-Bob beat cancer, and kicked his opioid addiction, and went on to live a long and happy life. The end.

But that is not what happened to countless other people. In fact, even the director of the CDC claimed "we know of no other medication [...] that kills patients so frequently" [1].

The thing is, Purdue Pharma knew about these issues all along [2]. It was clear in their clinical trials that their pills did not last 12 hours, as stated on the tin. They even told their prescribing physicians to increase patients' dosage instead of giving more doses a day, as this would help keep OxyContin in the black (other competing drugs lasted much less than the 12-hour span of OxyContin, so the big push was to maintain this "two doses a day" regimen to float its much higher price). Of course, increased dosage means a higher chance of addiction, potential overdose, and death. Nonetheless, Purdue kept pushing OxyContin onto physicians like it was going out of style. Other companies needed to get creative to eat into Purdue's market share, as they wanted to get rich quickly too. So, by the mid-2000s every big pharmaceutical company had their own miracle drug as well.

One great example of this is in a movie called "Pain Hustlers" on Netflix. Check it out. While it is a fictional portrayal of the very much real-life opioid crisis, it will clearly demonstrate how all of this went down. The pharmaceutical companies would hire attractive young salespeople (known as pharma reps) to go door-to-door to physicians' offices and preach the benefits of their opiates. Sometimes, if the physicians were a bit reluctant, the reps would grease the wheels a little. A new Porsche 993? Why not? You want an all-expenses paid "business trip" to Hawaii? Sure! In fact, if you prescribe enough, we might as well put you on our payroll! But that's not even where these companies stopped. Doctors were hired to be "speakers" at conferences to convince other doctors to prescribe the meds that they were working for! No conflict of interest there at all! If a physician was not eager to work with the rep, they would even sneakily (by bringing flowers and chocolates to the receptionist, for instance) take the names and phone numbers from the check-in logs and try to privately contact patients (and tell them that THEIR drugs are the ones they should be on, and why is their physician not prescribing it!?). These opioid companies have even been found in bed with electronic health record companies (the ones storing all of your medical data). Basically, for a "nominal" fee, these companies would release relevant information on who has what diseases, and in which areas, so the pharma reps could target those people specifically. So, Hippocratic oath out the window, sole focus is now on sales, sales, sales!

This means that regardless of whether or not it was the right thing to do for the patients, pharmaceutical companies encouraged doctors to prescribe more and more opiates to more and more patients for more and more reasons (by the end, it didn't even matter what you had – small headache? Here, try some OxyContin!). Meanwhile, the drug makers were also busy with the second of their two-pronged attack, marketing campaigns to downplay the risk of addiction with their new 12-hour pills, and overplaying the need for patient comfort. All for higher prescription rates. Of course, all of this was highly morally, ethically, and legally sketchy.

And yes, eventually all the lies did catch up with Purdue Pharma et al. In both 2007 (and again in 2020, 'cause they didn't stop the first time around), Purdue pleaded guilty to criminal charges for knowingly making false claims about the risks of opioid use. The company was also the target of thousands of lawsuits accusing it of being a catalyst for the opioid epidemic through its deceptive marketing of OxyContin. In the 20-year period between 1999 and 2019, nearly a quarter-million people died from overdosing on prescription opioids [3]. And through this, ladies and gentlemen, Purdue netted about $35 billion in profits from OxyContin.

Purdue ended up going bankrupt in 2019. In fact, the owners, the Sackler family, had the foresight to withdraw $11 billion from the company, right before the shit hit the fan, so that they could run away and live life in the lap of luxury [3]. Well, that didn't exactly go to plan though, as they were still kind of liable for the whole shebang. So, they proposed a plan that would reorganize the company into a nonprofit dedicated to addressing the problems created by the opioid epidemic (essentially the problems they caused through their greed and negligence). And how would they fund all this? They would take $4.5 billion out of the $11 billion the family took out of the company in the first place! And in exchange? The Sackler family would be released from all liability. Congrats Purdue Pharma for a job well done. Chapeau.

So now, dear reader, I can see you sitting on the edge of your seat, reading this riveting tale, and thinking to yourself, "yeah, but what's this got to do with me?!" Well, firstly, I hope you can see that not all decisions that physicians make are based on what is best for you as a patient, but sometimes rather just sales needs. The ever-present and all-encompassing dollar sign. But more importantly, I want you to think about how these companies attacked the world to gain a following... by flooding the market with advertising, sales reps, and through hired (and morally questionable) medical spokespeople. So, what's missing in this equation that is so prevalent in today's world? Had this case happened today, what could they have used to their tremendous advantage? Drum roll... SOCIAL MEDIA.

We live in this world of TikTok this and Instagram that. Everything from panda bears farting to teenage girls doing break dance challenges. Oh yeah, and a metric shit ton of (once again morally questionable) medical advice. So, let's take the OxyContin example and compare it with what's happening with Ozempic today. For those of you that may not know, Ozempic is an injectable EpiPen-looking drug that was developed solely for use by adults with Type 2 diabetes for controlling the user's blood sugar. But guess what? That's not why its selling in droves and causing global shortages... It's because of the Ozempic craze on TikTok. In fact, on TikTok alone, the hashtag "#ozempic" already has over 600 million views and counting [4]. It's trending because one side effect of the drug is miraculous weight loss. Now, I wanted to list a name of celebrities here for you that have jumped on the Ozempic bandwagon, and funnily enough, I went to Google to put in "celebrities on Ozempic" (as you normally do). I have put the predictive results for you in the following image (Figure 5.1).

Of course, Google uses my location, my previous search history, and a bunch of other data to tailor these results to me (so my results are going to be slightly different than yours). But more than any of that, Google is crowdsourcing all of the previously searched terms and using those to predict what I may want to see. And, of course, you have undoubtedly seen number one and number two on the list. Beating out "Celebrities Medical

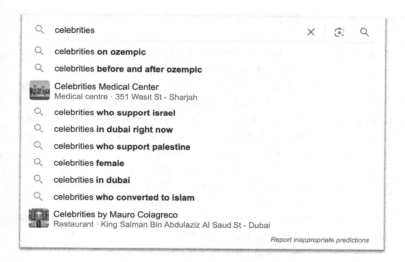

FIGURE 5.1 Google Predictive Search Results for the Term "Celebrities".

Center" and "celebrities who support Israel," we have our gold and silver medalists: "celebrities on Ozempic" and "celebrities before and after Ozempic." We have the likes of Oprah Winfrey, Kelly Clarkson, Amy Schumer, Whoopi Goldberg, Tracy Morgan (or is it Tracy Jordan, I always get those two confused), Charles Barkley. Pause for dramatic effect (and to take a deep breath). Now twice as fast… Billie Jean King, Khloe Kardashian, Al Roker, Rebel Wilson, Kate Winslet, Stephen Fry, Jessica Simpson, Mark Wahlberg, Chelsea Handler, Elon Musk, and the list goes on and on and on (I'm going to spare you the hundreds of others, you're welcome) [5, 6]. But the thing is, it's not just suspected that these celebs are all on Ozempic… they have all actively spoken out about their use of the magic drug on TV and social media. As Tracy Morgan said in an interview "I went and got a prescription, and I got Ozempic. And I ain't letting it go… It cuts my appetite in half. Now I only eat half a bag of Doritos." [5]. The problem is, are any of these people qualified to give medical advice to the masses? Do they have medical degrees? Fun fact- out of the list of names I just rattled off, not ONE has a medical degree, or a postgraduate degree of any kind (sorry Whoopi, but honorary degrees don't count). For example, Tracy Morgan, who I quoted just a moment ago, didn't even graduate from high school. As such, the authorities in the UK and Australia have even issued warnings to influencers promoting these drugs online, as pharmaceutical advertising is a highly regulated practice, which requires prior authorization [4].

So, now, we have a herd of medically unqualified celebrities talking about how great Ozempic is. Compounding this with the mountains of (possibly even less qualified) influencers on the social media platforms doing the same, and what do you think happens? People are heading to their family doctors and asking for Ozempic. Don't want to exercise or diet? Don't worry, someone is bound to prescribe Ozempic for you. And this is what's called "off-label" use (essentially just the use of pharmaceutical drugs for unapproved purposes or in an unapproved age group, dosage, way, etc.). So, this meteoric rise in people using Ozempic off-label is leading to a shortage, leaving the real target clientele (i.e., the patients who actually have diabetes) struggling to access it. In France, over the course of one year, 2,185 patients were given Ozempic, even though they weren't diabetic [4]. These kinds of numbers have forced countries like Australia and France to issue medical advisories for doctors to only prescribe the medicine to patients with Type 2 Diabetes WITH a risk of heart disease or stroke (thus acknowledging that there's a shortage and trying to prioritize those diabetics who need it most), and basically begging physicians to stop prescribing off-label [4].

Now, what's scary, is that no one seems to care, or even think about the side effects. Once again, this is not a miracle drug, and even Novo Nordisk, the creator of Ozempic says they do not "promote or endorse off-label use" of its products (and I'm sure they say this super seriously, with a very straight face, all while counting the stacks of money they're raking in) [4]. But in all seriousness, there are side effects. It's a range of stuff, from the likes of nausea and dehydration, all the way up to severe constipation or diarrhea, and inflammation of the pancreas [7]. In rare cases, even cancer [4]. And that's just the side effects that they have found in their medical trials. And who did they put in those medical trials? Type 2 diabetics (and definitely not a slightly overweight middle-aged woman from Ohio named Charlene who is just a tad lazy and has an appetite for double bacon cheeseburgers). So, what does this mean for us? Well, it means that Novo Nordisk, while testing the drug, did not test it for off-label use. Ergo, we don't actually even know what the full gamut of side effects can be for non-diabetics [8]. All I can tell you is, as of 2024 there have been 3000 reported cases of overdoses, as well as 100 deaths linked to Ozempic in the US, and another 20 in the UK, and one in Australia [9–12]. So, the big question is this: is a marginal amount of weight loss even worth a potential landmine of future health risks?

Well, the mass population have weighed in, and their answer is yes! It is worth it! But now, as doctors have begun cracking down and limiting the off-label prescriptions, how should you, the sweet, innocent, unsuspecting individual go about buying Ozempic? Well, you're in luck, there are still a few ways. One is to buy it directly from your good friend who does have a prescription! Guess what guys, diabetics are being prescribed Ozempic, and then turning around and selling it to the highest bidder on social media [13, 14]! There is even a huge black market dedicated to the illegal sale of prescription drugs online. BrandShield, a cybersecurity firm hired by a group of pharmaceutical companies, took down more than 250 of these websites in 2023, and they're saying it's only going to get worse [15]! In fact, things have gotten so bad (and naturally, all of it exacerbated by our good friend the internet), that people are now even selling FAKE Ozempic, and the WHO (the World Health Organization, not the band that sang Baba O'Riley and Pinball Wizard) had to even butt in and issue a warning on these "falsified medicines" [14–16]. So, rest assured, you won't be without your fix for long.

This, of course, is just one example of how you can get swept up in the world of medical information online. There are a ton more examples of influencers' genius medical advice going viral. Ozempic is simply just following on the heels of every other medical trend that came before it. All these diets. All these exercises. Botox. Fillers. Plastic Surgery. Every single one very real, and very current movements on social media [17, 18]. And all with very real side effects, too. Don't forget the Korean lady who was sued by her husband because their children popped out looking nothing like either of them, that's kind of like a side effect [19]. All in the name of beauty, and this idea of beauty all gathered from a series of Instagram reels and TikToks (Ian will go into more detail on these in his chapters). Now, going back to our Ozempic situation, just like the OxyContin scenario we discussed earlier, the negative side effects in both cases led directly to hospitalization, and even death. Now, imagine if we had the big three, that is TikTok, Instagram, and Facebook, in 1999. Imagine if #oxycontin began trending. How many more people would have died then? "So, what else could go wrong?" I hear you ask. Let's put on our scuba tanks, and dive into the murky waters of medical information and the internet.

Now, drifting slightly away from this idea (but still very much staying in the same lane) of not having the full picture before you recreationally decide to start taking a nice new pharmaceutical drug off-label, or having your face surgically pumped full of microplastics, all because your best friend's sister's neighbor's uncle's hairstylist said it was a great idea on Twitter, we also have this relevant topic of 'informed consent' that definitely needs to be discussed. Simply put, informed consent is the FULL disclosure of what could happen to a person if they were to, let's say, take a drug, undergo an operation, or just give any medical data to anyone. The full list of positives and negatives. Pros and cons. By giving this information to the patient or research subject, they can then go ahead and make a wise and informed decision about what to do (and do you think you're getting this from Mark Wahlberg's Instagram post?). For instance, if Dr. Wilkinson had told Joe-Bob, "listen Joe-Bob, this medicine will make you feel better in the short term, but in the long run, you will become addicted to it, and it's highly likely that you over-dose and die," I bet Joe-Bob would have probably told the doctor to shove that prescription right up his... clavicle.

But it's not just about telling the patient (or subject) every-thing, it's about doing it in such a way that they understand it. Mert, ha nem érted amit mondok, akkor miért is mondom el neked? For the 1% of you that speak Hungarian, you get my point. For the other 99% of you that don't speak Hungarian, you also get my point, you just don't know it yet. You must be bewil-dered, looking at this page like I lost the plot. The point is, I told you something. You just didn't understand it. Legally speaking, I still told you. And if that is all that informed consent is, then I would have fulfilled my legal obligations and medical duties. So, when a doctor says your "idiosyncratic hemodynamic response is indicative of a paradoxical pseudocomorbidities profile, necessitating a holistic polychromatic therapeutic approach," you would also likely have no idea what's going on. Therefore, in order to truly have informed consent, a doctor must explain what's going on. What is wrong with you, potential side effects of medications, and so on. Because only if you understand some-thing, can you make a decision about it. And lastly, in order to have full informed consent, the person making the decision also needs to make it voluntarily. If I tell you all of the medical side effects of Ozempic or OxyContin, the potential positives and negatives, all in a way you can understand, but then hold a gun to your head and say take it or I'll pull the trigger, that's not exactly informed consent.

Now, originally consent was simple. It was always for a sin-gle thing, and with a specified timespan or a specific purpose. But, with the rise of big data, we now have gigantic biomedical data warehouses, and everything just gets a little cloudy. It becomes more and more difficult to see what the future uses and implications of subject data could be, for instance. Therefore, we are at a crossroads, where the difficulty in getting informed consent, according to the original definition, is getting harder and harder (and many companies are just simply choosing not to). There are even more challenges being placed on informed consent when it comes to the medical data being gathered. Yes, ideally, we all want our medical data to be private and anonymous. But, for instance, some data can never really be anonymized (we'll talk more about this later). This puts a whole slew of new rules on the idea of informed consent and brings about the need for new solutions on how we can secure it. And therein lies the problem. In order for us to understand how to deal with the modern issues of consent, it's important, as always, to figure out where the idea even came from. But before we go that far, let's look at this from a broader sense.

Not getting informed consent is what we call "medical misconduct." As international laws get stricter and stricter when it comes to pharmaceutical sales, companies need to run larger and larger clinical trials. But these take time and effort. So, very often, the best way forward for these companies is to cut some corners. For instance, when the COVID-19 pandemic hit, every pharmaceutical company in the world was out to make a vaccine. But, of course, there was a huge time crunch, first to market would make a killing. So, what to do? Companies like Covaxin, who couldn't come up with enough legitimate volunteers to participate in the Phase III trials, ended up using unsuspecting victims... the general public [20]. Relying on residents in India's slums, Bharat Biotech (the maker of Covaxin) misled vulnerable people into believing they were receiving a COVID vaccine, when in fact, they were administering untested vaccines (and to some extent, so they say, placebos, but that is uncertain). That may even be worse. Imagine that someone in a white lab coat tells you that you are going to be vaccinated. You are ecstatic, because all of this COVID-19 propaganda that is being spewed everywhere has instilled a certain level of fear into you. So, you are happy to finally have a vaccine! Hooray! But, no one told you that this was a trial. No one told you that what you received could be a placebo. Hell, you don't even know what a placebo is!

Now, this kind of thing is not new at all. This type of misconduct in informed consent has been around for over a century in general medical care, and for many many decades in medical research. There have been countless examples of insane ethical oversights and morally flawed decisions made by both medical practitioners as well as researchers (we will discuss some of these in a minute). Now this is not even the worst of it. Today's version of this modern informed consent makes things even more bizarre. Giant leaps ahead in data gathering methods and essentially never-ending long-term storage options (with the now nearly infinite database capacities, no one ever needs to delete patient data again! They can just re-use it continually for newer and newer studies without your approval!). So, all these advances in research methodologies and new technologies. Do you think international regulations about informed consent have changed at all? Of course not!

Informed consent still relies heavily on these ancient legal statutes such as the 1946 Nuremberg Code, the 1964 Declaration of Helsinki, the 1979 Belmont Report, and the 1991 Common Rule (naturally, I should mention that the Declaration of Helsinki and the Common Rule have been revised, in 2013 and 2018 respectively, BUT they both still maintain the basic underpinnings of their original form) [21]. So, essentially, the idea of informed consent has totally eclipsed its classical form, however, the rules and regulations aren't updated enough to keep the misconduct at bay in this crazy, ever-evolving and data-hungry world of modern medical research [22]. So, let's take a look back in time and figure out why this is, and where it all came from.

Over the last century, the design of informed consent has come about from some truly dreadful events in two areas:

medical care, and human-subject (or medical) research. These two areas, of course, can be traced back far longer than a century, to the time of Hippocrates, and to the roots of western medicine (when medical practice was to conceal most information from patients, by the way), but it is the last 100 years that have been the most horrendous.

ISSUES WITH INFORMED CONSENT IN MEDICAL PRACTICE

There has been a lot of crazy shit done by physicians over the years. Maybe it's because of their authoritarian (and perhaps over-protective) approach to patient treatment (after all, the doctor is smart, and the patient is stupid). Despite rules dating all the way back to 1665, where even back then physicians and surgeons had to obtain a patient's consent prior to treatment under the Duke of York's laws of 1665 (the founding laws for New York), there have still been millions of court cases about unexpected results of medical procedures that patients opposed, did not approve, or were not provided with adequate information about [22]. Remember when we discussed the components of informed consent earlier? It's not just about you signing your name to a paper agreeing to a medicine or procedure, it's also about your doctor explaining everything related to that medicine or procedure in a way that you understand! So, keep that in mind when we talk about these next cases.

The first major court case that we're going to talk about is the 1905 case of *Pratt v. Davis*, where a surgeon removed a woman's uterus and ovaries, without ever even consulting her, in order to treat her epilepsy. Sound like a reasonable tradeoff? Well, here, the court ruled that a "physician or surgeon, however skillful [...] [cannot] violate without permission the bodily integrity of his patient" [23]. This step was one of the first major building blocks toward the modern version of informed consent. Another big one was the 1914 case of Schloendorff v. Society of

New York Hospital, where despite a patient's EXPLICIT refusal to a procedure, the surgeons still followed through with a surgery [24]. To make matters even worse, things didn't go exactly to plan, and doctors had to amputate some of the patient's fingers as a result. In this case, the court ruling was that "every human being of adult years and sound mind has a right to determine what shall be done with his own body" [24]. Another big step in cementing today's notion of informed consent (keep in mind the line about sound mind, we'll talk about that in a bit). The third notable court case (and last that we'll talk about today, so you can take a deep breath, we're almost done here) is the 1957 case of *Salgo v. Leland Standford Jr. University Board of Trustees*. Here, the term "informed consent" was used for the first time. A patient awoke permanently paralyzed after a routine procedure, and guess what, the patient had never even been informed that such a risk could exist. As such, the ruling of this case basically made it well-known that failure to disclose medical risks or alternative treatments could be cause for legal action [25].

SAME SAME, BUT IN MEDICAL RESEARCH

So, it was not just a physician's screw-up that caused informed consent to become a thing. When we talk about medical research, I would say that this section is way, way worse. This is the kind of stuff that will haunt your dreams. The fact that any of this ever happened is insane, to begin with.

Let's jump right in, and pretty deep from the very beginning... the Nazi's medical experimentation throughout World War II. Nazi doctors notoriously tortured and killed concentration camp prisoners in the name of scientific research. In 1946, these acts were heard in a trial known as "The Doctor's Trial" (the first of 12 trials against Germany for war crimes). Twenty of the 23 defendants were medical doctors. All were found guilty in the involvement of human experimentation and mass murder [26]. Ultimately, this ended up leading to the introduction of

The Nuremberg Code, and much later, indirectly led to the creation of the Declaration of Helsinki. Another famous example that arose due to the lack of proper research oversight was the Stanford Prison Experiment of 1971 [27]. A group of students were asked to act as prisoners and guards in an impromptu prison environment. Ultimately, they took their roles a little bit too seriously, resulting in prolonged psychological trauma for the subjects. The third and last (and arguably the most notoriously unethical human-subject research study in the history of the great US of A), is the Tuskegee Syphilis Experiment. This one's a big one. 600 underprivileged African Americans were studied regarding the effects of untreated syphilis over a 40-year span. However, they were originally told the experiment would last only a few months. The subjects were never even informed that they actually had syphilis either, and were not offered any medicine (despite there being a known cure) [28]. This, along with the Stanford Prison Experiment, led to the creation of the US Federal Policy for the Protection of Human Subjects (also known as the Common Rule), as well as the 1979 Belmont Report. Both of these, in essence, now require oversight processes for research involving human subjects by what we call "Institutional Review Boards" or "Ethical Review Boards." These are present in nearly every university and company doing research around the world.

Ok, so now that we know why we have informed consent, let's talk a bit about why it needs a bit of an update. As we mentioned before, informed consent needs to have three things [22]:

1. Information: Doctors and physicians (and anyone giving medical advice in general) need to disclose to you all the information related to a drug, a procedure, giving your medical data, etc., INCLUDING any and all risks, regardless of the effect that this may have on your willingness do that surgery, or take those certain medications.

2. Comprehension: All of the above information should be told to you in such a way that you can fully understand the information. The key here is that the person giving the information needs to understand the mental capacity of who they're dealing with (I hinted at this when we talked about Schloendorff v. Society of New York Hospital a few pages ago).

3. Voluntary participation: Of course, your consent has to be free and voluntary (both for entering and withdrawing from whatever it is that you are doing), with no caveats, coercion or influence from anyone. More on this later!

Now, these three aspects used to be easily addressable in the old-school way of informed consent gathering. However, today's medical research landscape has evolved a little bit since then, and with it, so has informed consent gathering. Quick example – you decide to hop on Ancestry.com to find out if you're 12% Lithuanian, because Aunt Angelika told you at the last family reunion that she grew up in Vilnius. In such a situation, consent is given by you clicking an itty bitty box underneath a tediously long online form. Does anyone even read these forms? Probably not. So, even if you clicked the "I Agree" box, do you really? As an aside, I don't know if you have seen, but some companies like Apple are now forcing you to scroll all the way through these contracts, before even allowing you to click the "Yes" button. That's their way of trying to get you to read the terms and conditions, but once again, does anyone really do it? So, does Ancestry.com actually have your consent to run your DNA? According to all that we talked about, and those three components of informed consent above, no ma'am, they certainly do not.

Alright, I see you sitting there, with your eyes glazing over like a Krispy Kreme donut from reading all of this medical mumbo jumbo. I know you're begging for me to tell you why on

God's green Earth this all has any bearing on your life. "I don't give out my medical data, so why do I care?" Well, the funny thing is, more people take part in giving their medical data away than you can imagine. In my house alone, there are 3 Macs, 2 iPads, 3 iPhones, 2 Apple Watches, 2 heart rate monitors (one synced up to a treadmill running iFit, the other to a bike with a Garmin Edge 1030 cycling computer), and a Garmin Index smart scale. And guess what? Every single one of these devices is able to gather, record, and store medical information of some sort and even transmits it to the great beyond (the internet). Thus, when setting up every single one of these devices, I had to acknowledge my giving informed consent, so that this specific doohickey can do its thing. Do you think I read ANY of those informed consent documents? Hell naw! I scrolled right past the pages and pages of text, straight to that big ol' "YES" button faster than a fly can find a pile of dog shit in the middle of an empty field under the noonday sun. And I imagine all of you do the same. Then, we go to the doctor's office, and give them access to our online-stored medical data as well by signing that little consent form at the front desk. The same goes for our health insurance… they have to get access to our records in order for them to pay for stuff, right? And what about the pharmacy? They keep your data too! We end up giving away far more medical data than we are aware of. And all of these are stored in a nice little secure server somewhere, kept safe under lock and key. Until they aren't. So, our lives are being monitored. Data is being recorded. But, our consent has been given! So, what's the issue? Well, to put it mildly, even if we read the entire 100-page consent form, it's highly likely that these companies may not have told you the truth, the whole truth, and nothing but the truth. Let me just count the ways in how these companies (the likes of Facebook, Google, FitBit, and Apple, just to name a few) have lied to us.

MEDICAL DATA GATHERED OR SHARED WITHOUT CONSENT

So, let's start with the most blatant disregard of informed consent – companies gathering medical data without even telling their users. Like the time the University of Toronto managed to get their grubby little paws on the electronic medical records of over 613,000 people in 2023 [29]. They allegedly conned family doctors to submit entire patient charts under the ruse of needing it for a research study. However, there was no research study, and as a whistleblower inside the university pointed out, it was all gathered to form one giant server of medical data, that they could then turn around and sell to third parties. Naturally, the University of Toronto has denied all of these allegations, and declined to address the concerns of the public. But at the same time they also announced that they would pause the collection, use, and transfer of any of this data… interesting. They don't sound guilty at all… I'm perfectly sure that lots of innocent people decline to comment in the public arena about allegations against them and immediately say they will stop doing what they were accused of but weren't doing. "I swear officer, I wasn't selling crystal meth. No, I can't go to the station with you right now, but I promise that I will stop selling crystal meth." Makes perfect sense. The whistleblower went on in their complaint that "patients were not afforded any real opportunity to withdraw from participation" and that they "were completely unaware (and remain unaware) that this was even happening" [29]. A blatant and direct violation of informed consent. But it was not just the University of Toronto to blame in this example, don't forget to blame the physicians that also handed over those medical records (most likely) against the informed consent that their patients had given them. I'm sure no one volunteered their medical data to be given away by their family physician to a series of undisclosed recipients.

But this kind of stuff happens all of the time. Facebook has been caught monitoring and receiving patient information from hospital websites through users' browsers [30]. Google keeps track of your health-related internet searches [30]. Health apps leave room in their privacy policies to share data with unlisted third parties [30]. For instance, the apps Drugs.com Medication Guide, WebMD: Symptom Checker, and Period Calendar Period Tracker have all been guilty of giving their advertisers users' medical information. To dig a little deeper here, these apps basically tracked terms like "herpes," "HIV," "Adderall," "diabetes," "addiction," "depression," and "pregnancy," and sent them over to the advertisers, so that they can target users with products based on their specific health concerns [30]. Did anyone even give permission for this? Of course not! But, ambiguously stated consent forms are allowing this stuff to happen. So, once again, all three categories of informed consent cannot be achieved. If people are given one million-page legal document they can't understand, they have no idea what's happening to their data, and all the while, they can't even opt out from all this, then information, comprehension, and voluntariness are all null and void.

To make matters even scarier, companies are now turning your everyday data INTO medical data (also without your consent). One great example of this was when Ted Cruz ran against Donald Trump for the Republican party's nomination for President of these here United States of America (cue eagle screech). It turned out that he was harvesting psychological data gathered from tens of millions of Facebook users, all without their consent [31]. In a nutshell, they used a company called Cambridge Analytica to basically access a large pool of Facebook profiles, and grab tens of thousands of people's demographic data (such as names, locations, birthdays, genders, etc.), as well as their Facebook "likes" [31]. They then used this data to categorize people into personality groups using the so-called "big five" personality traits known as the OCEAN scale (openness, conscientiousness, extraversion, agreeableness, neuroticism)

[31]. This then offered a huge range of personal insights about the potential constituents, and Teddy boy was able to create highly targeted campaign messages for specific issues, and communicate them in multiple ways to different audiences depending on where he was and who he was talking to [31]. But wait! It gets even better! They used these personality traits to develop a series of TV ads, each aimed at different personality types, and they aired them at times when viewers with those personalities were most likely to be watching! Neat huh? Of course, Cruz's office said they had no idea the data was taken without consent ("my understanding is all the information is acquired legally and ethically with the permission of the users when they sign up to Facebook") [31]. Reminds me of when Richard Nixon said that famous line, "I am not a crook" with his cheeks all wobbling. Oh, and what about Cambridge Analytica? Guess what... just like a true innocent bystander, they refused to comment on the allegations and even hung up the phone on the reporter.

MEDICAL DATA NOT STORED SECURELY OR LEAKED

Now onto a horse of a different color. The second big issue we seem to run into with online medical information is security. Let's just hypothetically say that you have, in fact, consented to giving your medical data to 23andMe in order for them to run a background check on your ethnicity and learn more about your family tree. You send them that cheek swab, after all, they claim it's super easy – "It's just saliva. No blood. No needles" [32]. So, what could go wrong? Well, actually, quite a bit. In late 2023, a group of black-hat hackers gained access to all of this information. Out of 23andMe's 14 million users, they got into 6.9 million of them [33, 34]. Then they started targeting people. First, it was the Jews. Then it was the Chinese. The hackers ended up releasing all of the files of people's data, grouped by these two categories. The data profiles were also made available for purchase on the black market [33]. As this just happened a few months ago, at the time of writing this, no one knows what can come of this. But I'm pretty sure it ain't gonna be good. Naturally, there is a big concern about this leaked information leading to targeted attacks. Immediately following this leak, 23andMe's two main competitors, Ancestry.com and MyHeritage, began requiring users to utilize two-factor authentication [35]. Coincidence? I think not. 23andMe was sued for negligence, breach of implied contract, and invasion of privacy among other charges [36]. The case is still pending, but in December of 2023, 23andMe updated their terms and conditions to prevent any class action lawsuits [37]. So even if they leak out all of your data, there's nothing you can do about it. Not shady at all!

Once again, this is not the first time such a leak has happened. 61 million fitness tracker records were exposed from the likes of Apple and FitBit in 2021 [38]. A company called GetHealth was storing all of their medical data from fitness trackers and wearable devices in a non-password-protected database. The records contained user data including first and last name, username, date of birth, weight, height, gender and geolocation, and while the data was mostly from Fitbit devices and Apple Healthkit, they also sync health-related data from other sources, including Google Fit, Jawbone UP, Microsoft, Sony Lifelog, Withings, Android Sensor, and guess who else... 23andMe (not you guys again!) [38]. Another similar situation happened in Google a few years prior. Google was running a secret project transferring the personal medical data of 50 million Americans from Ascension, the second largest healthcare provider in the US, to Google [39]. This so-called "Project Nightingale" collected medical data from 21 states, and no one even knew about it. This was by far the largest data transfer of its kind in the healthcare field. And guess what? The data was transferred with full personal details of all of the patients, including name and medical history. Wanna know even more? It could be accessed by any of Google's staff. No effort was made to secure or anonymize any of the data [40].

Thankfully, nothing seems to have leaked, but there was still that one very pissed-off employee that revealed all of this to the world. And that doesn't exactly make Google look good.

And that brings me to my next, and thankfully, last point. This idea of anonymization, and whether it actually means or does anything. For those of you that may not know, anonymization is the process of removing personal information from medical data. According to modern practices, most people and companies seem to think that this is sufficient to secure people's medical data (and usually, they even tell you this in your informed consent forms so that you are more willing to give your data). But is this really the case? Let me tell you a little story. In the mid-1990s, the Massachusetts Group Insurance Commission had a bright idea to release anonymized medical data on state employees [41]. The idea behind this all was to help researchers gain access to such medical data in order to run experiments, develop trends, predict the future, take over the world, etc. Researchy stuff. So, the state spent a boatload of time removing all of the obvious identifiers in the medical histories, such as name, address, and Social Security numbers. Perfect. You're welcome researchers! Well, one computer science student by the name of Latanya Sweeney thought this was a stupid idea, and decided to prove it. At the time the data was released, William Weld, the Governor of Massachusetts, assured everyone that the state had protected patient privacy by deleting all of these identifiers. So, Sweeney went fishing for the Governor's hospital records. She had a few key pieces of information about him, like that he lived in Cambridge, Massachusetts (which, conveniently only had 54,000 residents and seven ZIP codes). For twenty bucks, she purchased the complete voter records for Cambridge, which, you guessed it, held the name, address, ZIP code, birth date, and sex of every registered voter in the city. By combining this data with the "anonymous" state medical records (I use parentheses because we both know they weren't truly anonymous), Sweeney found the Governor with relative ease, as only six people in Cambridge shared his birth date, only three of them were men, and of these men, only he lived in that specific ZIP code [41]. Once she got this far, she sent the Governor his own health records, including his recent diagnoses and prescriptions. I hope Mr. Weld had his EpiPen ready to open up his airways, because I'm sure he choked on that. But, it wasn't just his medical records that could be found. Sweeney showed that 87% of all Americans could be uniquely identified using only three pieces of information: **ZIP code, birthdate, and gender** [41]. Damn!

Essentially, since then, we've learned that "personal information," even if anonymized, can be made "personal" again. All you need is the right bits of data (once again, pun intended). So why is it that time and time again, people want to revisit this idea of anonymization? Just in 2014, NHS, the UK's public healthcare system wanted to put all UK medical data into a public database [42]. Tim Kelsey, NHS England's National Director for Patients and Information even went so far as to say "Can I be categorical? No one who uses this data will know who you are." [42]. Keep in mind this is two decades after Latanya Sweeney called the Massachusetts Governor on his bullshit. And this goes on and on. While Sweeney said she could identify 87% of Americans with three characteristics, scientists are now saying that they can identify 99.98% with 15 characteristics. And it's not just medical data… this is exactly how Donald Trump's tax returns ended up going public in 2019 [43]. In 2014 the home addresses and incomes of all New York cabbies were uncovered from a huge set of anonymized data (the city released detailed data on 173 million taxi trips) [44]. Going back a little further, in 2008, an anonymized Netflix dataset of half a million users' film ratings was deanonymized by comparing the ratings against public scores on the IMDb website [44]… And all of this is not stopping. Companies continue publishing and selling datasets, because that's what it all comes down to. Money. Companies need data. Researchers need data. Politicians need data. And they're all willing to pay big bucks for it. So, let me wrap this up by putting another log on the fire. Firms like Experian sell heaps of these anonymized datasets. And they contain infinitely more information per person than you think... one such dataset that was recently sold, and included anonymized data for 120 million Americans. Now, remember how I told you that 99.98% of Americans could be identified with just 15 characteristics? Experian's datasets held 248 [45].

So, what have we learned from all of this? Let's sum it up as easily as possible. These are the things that YOU should be most weary about when dealing with medical data and information (be it you giving your data, your consent, or just getting medical information off of a sketchy source like TikTok). Once again, it all ties back to the information, comprehension, and voluntariness trilogy, and by now, you should know that like the back of your hand.

When we talk about information, think about not just where the information is coming from, and the validity that it has, but also keep in mind the outgoing side, the data sharing. Information is a two-way street. You need to know everything to give consent (it's not enough to just read Elon Musk's tweets about Ozempic, and decide that's the thing for you). On the flip side, whenever giving your medical data to anyone, you need to think, what happens if this information gets shared with others? What value may it have to someone, and why would they want it in the first place (what do they stand to gain by having it – think of all the companies that make a killing on sharing private data for marketing purposes, for instance)? Second, of course, you can be overwhelmed by the sheer complexity or volume of information that is coming at you in the medical world (and the tangential world of medical information on social media). If you cannot understand the necessary details, be like the opposite of Nike. DON'T DO IT (whatever "it" may be). And lastly, make your own decisions, and make sure you do them voluntarily. Pay special heed to those that are more vulnerable (for example, children, the elderly, the mentally impaired, etc.). It becomes doubly important to protect them, and ensure that they're not getting themselves into sketchy situations. So, keep an eye out for your little sister when she wants to get those new-fangled lip fillers because Kylie Jenner looked so hot on her last Instagram reel. If you don't look out for her (and explain to her the three categories of informed consent), she may just end up becoming the next new star of Botched. As Lewis Armstrong so eloquently said, "what a wonderful world."

REFERENCES

[1] S. Moghe, "Opioid history: From 'wonder drug' to abuse epidemic," CNN, Oct. 14, 2016. [Online]. Available: https://edition.cnn.com/2016/05/12/health/opioid-addiction-history/index.html. [Accessed: Jun. 21, 2024].

[2] H. Ryan, L. Girion and S. Glover, "'You want a description of hell?' Oxycontin's 12-hour problem," Los Angeles Times, May 5, 2016. [Online]. Available: https://www.latimes.com/projects/oxycontin-part1/. [Accessed: Jun. 21, 2024].

[3] A. Howe, "Justices put Purdue Pharma bankruptcy plan on hold," SCOTUSblog, Aug. 10, 2023. [Online]. Available: https://www.scotusblog.com/2023/08/justices-put-purdue-pharma-bankruptcy-plan-on-hold/#:~:text=OxyContin%20first%20came%20on%20the,in%20revenue%20for%20Purdue%20Pharma. [Accessed: Jun. 21, 2024].

[4] O. Duboust, "Ozempic: How a TikTok weight loss trend caused a global diabetes drug shortage – and health concerns," Euronews, Mar. 3, 2023. [Online]. Available: https://www.euronews.com/health/2023/03/02/ozempic-how-a-tiktok-weight-loss-trend-caused-a-global-diabetes-drug-shortage-and-health-c. [Accessed: Jun. 21, 2024].

[5] K. Hogan, A. Schonfeld, T. Parent and V. Verbalaitis, "Stars who've spoken about Ozempic – and What they've said," People, Jun. 10, 2024. [Online]. Available: https://people.com/celebrities-ozempic-wegovy-what-theyve-said-7104926. [Accessed: Jun. 21, 2024].

[6] C. Callahan, "18 celebrities who've opened up about taking Ozempic or weight-loss drugs," Today, May 21, 2024. [Online]. Available: https://www.today.com/health/celebrities-on-ozempic-rcna129740. [Accessed: Jun. 21, 2024].

[7] A. Northrop, "Ozempic for weight loss: Is it safe? What experts say," Forbes, Jun. 7, 2024. [Online]. Available: https://www.forbes.com/health/weight-loss/ozempic-for-weight-loss/. [Accessed: Jun. 21, 2024].

[8] D. Blum, "What Is Ozempic and Why Is It Getting So Much Attention?," The New York Times, Nov. 13, 2023. [Online]. Available: https://www.nytimes.com/2022/11/22/well/ozempic-diabetes-weight-loss.html. [Accessed: Jun. 21, 2024].

[9] J. Childs, "Ozempic overdose? Poison control experts explain why thousands OD'd this year," Los Angeles Times, Dec. 20, 2023. [Online]. Available: https://www.latimes.com/science/story/2023-12-20/semaglutide-ozempic-wegovy-overdoses. [Accessed: Jun. 21, 2024].

[10] L. Andrews, "More than 100 US deaths linked to Ozempic and similar weight loss drugs - including 28-year-old who died from 'intestinal mass' and a pregnant woman, our analysis shows," Daily Mail, Apr. 11, 2024. [Online]. Available: https://www.dailymail.co.uk/health/article-13276757/deaths-linked-ozempic-weight-loss-drugs-analysis.html. [Accessed: Jun. 21, 2024].

[11] E. Stearn, "Revealed: Weight loss jabs like Ozempic have been linked to TWENTY deaths in Britain - including person in their 30s - as experts issue new warning," Daily Mail, Apr. 5, 2024. [Online]. Available: https://www.dailymail.co.uk/health/article-13266925/Weight-loss-jabs-like-Ozempic-linked-TWENTY-deaths-Britain.html. [Accessed: Jun. 21, 2024].

[12] G. Chung, "Mom dies after taking Ozempic to lose weight for daughter's wedding," NBC Miami, Nov. 10, 2023. [Online]. Available: https://www.nbcmiami.com/news/national-international/mom-dies-after-taking-ozempic-to-lose-weight-for-daughters-wedding/3156998/. [Accessed: Jun. 21, 2024].

[13] A. Walsh and P. Rai, "Weight loss injection hype fuels online black market," BBC News, Nov. 15, 2023. [Online]. Available: https://www.bbc.com/news/health-67414203. [Accessed: Jun. 21, 2024].

[14] G. Mammoser, "TikTok influencer accused of selling fake Ozempic and other weight loss drugs," Healthline, May 9, 2024. [Online]. Available: https://www.healthline.com/health-news/tiktok-influencer-sold-fake-ozempic#:~:text=A%20 New%20York%20TikTok%20influencer,faced%20"life%20 threatening"%20injuries. [Accessed: Jun. 21, 2024].

[15] S. Goodyear, "Hundreds of websites are selling fake Ozempic, says company. Doctors say it's only going to get worse," CBC Radio, Apr. 19, 2024. [Online]. Available: https://www.cbc.ca/ radio/asithappens/fake-ozempic-report-1.7176597. [Accessed: Jun. 21, 2024].

[16] "WHO issues warning on falsified medicines used for diabetes treatment and weight loss," World Health Organization, Jun. 20, 2024. [Online]. Available: https://www.who.int/news/ item/20-06-2024-who-issues-warning-on-falsified-medicines-used-for-diabetes-treatment-and-weight-loss. [Accessed: Jun. 23, 2024].

[17] E. Upton-Clark, "Gen Z's new status symbol," Business Insider, May 2, 2024. [Online]. Available: https://www.businessinsider. com/botox-injections-fillers-plastic-surgery-gen-z-social-media-filters-2024-5. [Accessed: Jun. 21, 2024].

[18] K. Peterson, "Navigating social media's influence on plastic surgery decisions," Plastic Surgery, Jun. 29, 2023. [Online]. Available: https://www.plasticsurgery.org/news/ articles/navigating-social-medias-influence-on-plastic-surgery-decisions#:~:text=In%20recent%20years%2C%20 the%20impact,showcasing%20before%20and%20after%20 transformations. [Accessed: Jun. 21, 2024].

[19] "Can't hide it forever': The model who became a meme," BBC News, Oct. 29, 2015. [Online]. Available: https://www.bbc. com/news/world-asia-34568674. [Accessed: Jun. 21, 2024].

[20] E. Mitra and J. Hollingsworth, "More than a dozen slum residents in an Indian city say they thought they were being vaccinated. They were part of clinical trials," CNN, Feb. 26, 2021. [Online]. Available: https://edition.cnn.com/2021/02/25/ asia/india-vaccine-trials-covid-ethics-intl-dst-hnk/index.html. [Accessed: Jun. 21, 2024].

[21] B. Cassidy, "Medical research is changing, data privacy laws have not," Healthcare Dive, Apr. 30, 2024. [Online]. Available: https://www.healthcaredive.com/news/medical-research-data-privacy-laws-hipaa-senator-bill-cassidy/714609/. [Accessed: Jun. 21, 2024].

[22] M. Gergely, F. K. Dankar, and S. Alrabaee, "Misconduct and consent: The importance of informed consent in medical research," in *Integrity of Scientific Research: Fraud, Misconduct and Fake News in the Academic, Medical and Social Environment*, J. Faintuch and S. Faintuch, Eds. Cham: Springer, 2022.

[23] L. A. Bazzano, J. Durant, and P. R. Brantley, "A modern history of informed consent and the role of key information," *Ochsner Journal*, vol. 21, no. 1, pp. 81–85, Spring 2021.

[24] "Schloendorff v. New York Hospital," CaseText, Apr. 14, 1914. [Online]. Available: https://casetext.com/case/schloendorff-v-new-york-hospital. [Accessed: Jun. 21, 2024].

[25] "Salgo v. Leland Stanford Jr University Board of Trustees (1957)," FindLaw, Oct. 22, 1957. [Online]. Available: https://caselaw.findlaw.com/court/ca-court-of-appeal/1759823.html. [Accessed: Jun. 21, 2024].

[26] "Doctors' Trial," Wikipedia. [Online]. Available: https://en.wikipedia.org/wiki/Doctors%27_Trial. [Accessed: Jun. 21, 2024].

[27] "Stanford prison experiment," Wikipedia. [Online]. Available: https://en.wikipedia.org/wiki/Stanford_prison_experiment. [Accessed: Jun. 21, 2024].

[28] "The USPHS Untreated Syphilis Study at Tuskegee," CDC, Jan. 9, 2023. [Online]. Available: https://www.cdc.gov/tuskegee/index.html#:~:text=The%20U.S.%20Public%20Health%20Service%20(USPHS)%20Untreated%20Syphilis%20Study%20at,after%20it%20was%20widely%20available. [Accessed: Jun. 21, 2024].

[29] A. Russel and C. Lieberman, "Whistleblowers allege U of T data project collected 600K patient records without consent," Global News, Jan. 24, 2023. [Online]. Available: https://globalnews.ca/news/9428080/university-of-toronto-medical-records-data-project-ontario-privacy-complaint/. [Accessed: Jun. 21, 2024].

[30] T. Hunter and J. B. Merrill, "Health apps share your concerns with advertisers. HIPAA can't stop it," Washington Post, Sep. 22, 2022. [Online]. Available: https://www.washingtonpost.com/technology/2022/09/22/health-apps-privacy/. [Accessed: Jun. 21, 2024].

[31] H. Davies, "Ted Cruz using firm that harvested data on millions of unwitting Facebook users," The Guardian, Dec. 11, 2015. [Online]. Available: https://www.theguardian.com/us-news/2015/dec/11/senator-ted-cruz-president-campaign-facebook-user-data. [Accessed: Jun. 21, 2024].

[32] "How 23andMe works," 23andMe. [Online]. Available: https://www.23andme.com/en-int/howitworks/#. [Accessed: Jun. 21, 2024].

[33] "23andMe data leak," Wikipedia. [Online]. Available: https://en.wikipedia.org/wiki/23andMe_data_leak. [Accessed: Jun. 21, 2024].

[34] L. H. Newman, "23andMe User Data Stolen in Targeted Attack on Ashkenazi Jews," Wired, Oct. 6, 2023. [Online]. Available: https://www.wired.com/story/23andme-credential-stuffing-data-stolen/. [Accessed: Jun. 21, 2024].

[35] A. Belanger, "After hack, 23andMe gives users 30 days to opt out of class-action waiver," Ars Technica, Dec. 7, 2023. [Online]. Available: https://arstechnica.com/tech-policy/2023/12/23andme-changes-arbitration-terms-after-hack-impacting-millions/. [Accessed: Jun. 21, 2024].

[36] "Santana et al v. 23andMe, Inc.," Bloomberg Law, Oct. 9, 2023. [Online]. Available: https://www.bloomberglaw.com/public/desktop/document/Santanaetalv23andMeIncDocketNo323cv05147NDCalOct092023CourtDocket?doc_id=X1Q6OKJP4D82. [Accessed: Jun. 21, 2024].

[37] Z. Whittaker, "23andMe data theft prompts DNA testing companies to switch on 2FA by default," TechCrunch, Nov. 7, 2023. [Online]. Available: https://techcrunch.com/2023/11/07/23andme-ancestry-myheritage-two-factor-by-default/. [Accessed: Jun. 21, 2024].

[38] H. Landi, "Fitbit, Apple user data exposed in breach impacting 61M fitness tracker records," Fierce Healthcare, Sep. 13, 2021. [Online]. Available: https://www.fiercehealthcare.com/digital-health/fitbit-apple-user-data-exposed-breach-impacting-61m-fitness-tracker-records. [Accessed: Jun. 21, 2024].

[39] R. Copeland, "Google's secret project Nightingale gathers personal health data on millions of Americans," Wall Street Journal, Nov. 11, 2019. [Online]. Available: https://www. wsj.com/articles/google-s-secret-project-nightingale-gathers-personal-health-data-on-millions-of-americans-11573496790. [Accessed: Jun. 21, 2024].

[40] E. Pilkington, "Google's secret cache of medical data includes names and full details of millions – whistleblower," The Guardian, Nov. 12, 2019. [Online]. Available: https://www. theguardian.com/technology/2019/nov/12/google-medical-data-project-nightingale-secret-transfer-us-health-information. [Accessed: Jun. 21, 2024].

[41] N. Anderson, "'Anonymized' data really isn't—and here's why not," Ars Technica, Sep. 8, 2009. [Online]. Available: https://arstechnica.com/tech-policy/2009/09/your-secrets-live-online-in-databases-of-ruin/. [Accessed: Jun. 21, 2024].

[42] E. Bolten, "Your NHS data is completely anonymous – until it isn't," The Conversation, Feb. 10, 2014. [Online]. Available: https://theconversation.com/your-nhs-data-is-completely-anonymous-until-it-isnt-22924. [Accessed: Jun. 21, 2024].

[43] M. Yglesias, "The battle over Trump's tax returns, explained," Vox, May 10, 2019. [Online]. Available: https://www.vox.com/policy-and-politics/2019/5/10/18537282/trump-tax-returns-congress-irs. [Accessed: Jun. 21, 2024].

[44] A. Hern, "New York taxi details can be extracted from ano-nymised data, researchers say," The Guardian, Jun. 27, 2014. [Online]. Available: https://www.theguardian.com/technology/2014/jun/27/new-york-taxi-details-anonymised-data-researchers-warn. [Accessed: Jun. 21, 2024].

[45] A. Hern, "'Anonymised' data can never be totally anonymous, says study," The Guardian, Jul. 23, 2019. [Online]. Available: https://www.theguardian.com/technology/2019/jul/23/anonymised-data-never-be-anonymous-enough-study-finds. [Accessed: Jun. 21, 2024].

6 I Spy with My Little Eye – The Internet and Digital Espionage

America is a vast conspiracy to make you happy.
—John Updike

2004 – Enlists to serve in the US Army, specifically the Army Reserve Special Forces, breaks both of his legs in a training exercise (or so he claims). Receives administrative discharge and is encouraged to help serve his country in other ways [1, 2].

2006 – Undergoes a screening process for a job at the CIA. Despite having insufficient answers, the Deputy Director decides to take a chance on him. Then brought to "The Hill" (Station B of Warrenton Training Center, in Virginia), and educated on cyberwarfare. Given top-secret clearance, and learns about the Foreign Intelligence Surveillance Act, which circumvents the Fourth Amendment rights of civilians (which governs the search and seizure of property) [3].

2007 – Posted in Geneva, Switzerland, under diplomatic cover as an IT and cybersecurity specialist, giving him access to a wide array of classified documents. Begins questioning the ethicality of the actions of the US government. After his supervisor sets up their target on a bogus DUI charge to coerce information out of him, resigns from the CIA [1].

DOI: 10.1201/9781032679389-7

2009 – Takes a position with the NSA in Japan, tasked to work on a project he named "Epic Shelter", building a computer program that would allow the government to back up critical data from the Middle East in the case of an emergency. Quickly learns that the NSA and other US government agencies are injecting malware into different computer systems managing government, infrastructure, and financial sectors in allied countries, so in the event that any allies turn against the US, that country can effectively be neutralized via these malwares [3].

2012 – Moves to Hawaii to work as a contractor for Dell at "The Tunnel", an NSA facility in an underground WWII bunker that has been re-purposed for electronic surveillance and signals intelligence operations (intercepting and analyzing communications). Learns that the program he was working on, Epic Shelter, is actually being used to provide real-time data to assist US drone pilots in launching deadly strikes against terrorism suspects [1, 3].

April 9, 2013 – Gets hired by Booz Allen Hamilton (a US government contractor specializing in intelligence) in order to covertly gain access to top-secret information that he intends to leak and make public. Begins sending confidential US-government documents to journalists [1, 3].

May 20, 2013 – Tells his NSA supervisor that he needs to take some time off for medical treatment. Moves to Hong Kong, as the territory's press freedom offers many safeguards, and its extradition treaty with the US is favorable to him [1, 3].

June 2, 2013 – Reveals himself as a whistleblower and releases thousands of classified NSA documents to a few select journalists about the collection of domestic emails and telephone data from millions of Verizon customers (a large US telecom provider), which was just a small part of an even wider program called PRISM [3, 4].

June 5, 2013 – The Guardian and The Washington Times publish articles about how the US government agencies are forcing the largest telecom companies to hand over civilian users' private data [4].

June 10, 2013 – Terminated by Booz Allen Hamilton for breach of the firm's code of ethics and policy. The company releases a statement that "news reports that this individual has claimed to have leaked classified information are shocking, and if accurate, this action represents a grave violation of the code of conduct and core values of our firm" [5].

June 14, 2013 – Charged under the 1917 Espionage Act with theft, "unauthorized communication of national defense information" and "willful communication of classified communications intelligence information to an unauthorized person" by the US Justice Department [3].

June 23, 2013 – Leaves Hong Kong for Ecuador. Gets stranded in Russia during a layover after US authorities cancel his passport. Spends over a month living in the airport (à la The Terminal with Tom Hanks) [1, 3].

August 1, 2013 – Granted temporary asylum, while the Russian authorities process his application for permanent political asylum. At the same time, The Guardian publishes another article (based on the leaked information) about how the NSA has been paying off UK intelligence agencies to develop techniques that allow the mass harvesting and analysis of internet traffic in an effort to gather personal information from mobile phones and apps, to be able to "exploit any phone, anywhere, anytime" [6].

October 14, 2013 – Using the same documents, The Washington Post reveals that the NSA has used the fiber optic cables that carry most of the world's telephone and internet traffic to collect data from over 250 million accounts a year from the likes of Yahoo, Gmail and Facebook [7].

2014 – Receives three-year Russian residency permit after one-year temporary asylum expired [8].

2020 – Granted permanent residency in Russia [9].

2022 – Given Russian citizenship by President Vladimir Putin [10].

Now, the question I'm sure you're pondering is "who could we be possibly talking about here?" If you couldn't guess, it's Mr. Edward Snowden, of Elizabeth City, North Carolina. Played by Joseph Gordon-Levitt in the feature film Snowden, where I have to say, the casting was really, really good. They look like twins. Anyways, Edward Snowden revealed to the world how un-private your private data and communications are. To take a quote from Mark Twain "if voting made any difference, they wouldn't let us do it" (well, technically, it's unproven who actually said this, but hey, whatever, it's broadly attributed to Mr. Twain, so let's go with that). What does this have to do with Snowden? Well, in the same vein, if the government wants to access your private data, they will. This has not been the first time a government has been caught spying on its citizens, and you can be pretty damn sure it won't be the last.

Just immediately before the whole Snowden debacle was Julian Assange, another US enemy of the state (shout out to the old Will Smith movie, in case you haven't seen it). He also revealed classified government documents through his website WikiLeaks, said to be in the neighborhood of 100 million individual documents related to war, spying, and corruption [11]. The published documents were gathered by him hacking into government databases, as well as stuff submitted by ex-government employees (like Chelsea Manning, who used to work for the US Army). In 2019, the US Department of Justice called this "one of the largest compromises of classified information in the history of the United States".

So why are these leaks such a big problem? Well, firstly (from a military standpoint), they are said to compromise the whereabouts of agents in the field, giving away people's identities, revealing governmental and military plans, and so forth. Yes, clearly that's an issue. But, what was truly a stunning revelation, was the amount of PERSONAL data that the US government was holding (some would say illegally, or at least unethically) about private individuals (such as the daily recordings of millions of people's phone calls, and their keeping tabs on people's phone and internet records) [1]. You see, the world wasn't always like this. Let's hop back a few centuries and see why. During the Revolutionary War, the Americans were fighting the Brits. During this time, one thing really kept pissing Americans off, and that was that British soldiers kept coming into American households and just taking it over. They would take food, weapons, and just anything they wanted. This is not uncommon in communist countries, or those with a monarchy… technically everything belongs to the ruler or the government, and therefore, they can just take whatever they want. In addition, the soldiers of that ruler or government, are acting as agents FOR that ruler or government, and therefore, they can do the same thing. This is what we call a 'general warrant'. Of course, the Americans were all like "this is mine" and "that's bullshit", so they made a vow. If they were to win the war, they would make a law stating that private property cannot be simply taken away by the government. Hence, when in 1776 the Declaration of Independence was signed, it included the Fourth Amendment. This piece of legislation essentially protects people from having their belongings searched or seized by the government… "The right of the people to be secure in their persons, houses, papers, and effects, against unreasonable searches and seizures" [12]. So why are we sitting here discussing a 250-year-old law? Well, ladies and gentlemen, because modern-day surveillance laws are still based around this ancient statute. Did you see where they referred to "papers" in the quote above? Well, this is now your data. Your Tinder profile data, your WhatsApp messages, your Snapchat DMs, and even your phone conversations.

Over time, there were a few newer laws added to spruce up the legal landscape and add to the Fourth Amendment. One such example is the Electronic Communications Privacy Act of 1986, otherwise known as the Wiretap Act, that prohibited government officials to record phone conversations without a warrant. In general, not much changed. Certain actions were clarified as legal or illegal in keeping with the times, but the overall essence of the Fourth Amendment stayed the same – no searches and seizures of physical property or electronic data without a warrant. But that all went to shit in 2001. After 9/11, George Bush (the painter, not the sock collector) signed into law a little ditty called the Patriot Act. In a nutshell, it takes the Fourth Amendment, the Wiretap Act, and all similar laws, and rips them up. Now, all a government entity would have to say are a few special words, and they wouldn't need a warrant. Those magic words were "it's a matter of national security". Upon uttering those simple words, agencies now had comprehensive surveillance abilities, including tapping domestic and international phones, and internet-based communications [13, 14]. And of course, no one took advantage of this at all. Everyone was very ethical and lived happily ever after. The end. Sure… There were tons of cases where people suddenly were being charged for various crimes based on evidence that had suddenly appeared without a warrant. GPS tracking data placed on people's cars, home searches, recorded phone calls. However, all of this seemed rather limited in scope, until Ed Snowden revealed the magnitude of the whole enterprise.

In fact, the Patriot Act was so controversial that the ACLU (American Civil Liberties Union) claimed it "threatens our most basic civil liberties" [15]. Then how on Earth did this law get passed? Well, many claim it was on the back of events like 9/11, where any measure to make the US a safer place, seemed like a good idea (like taking off your shoes at an airport). Michael Moore, in his documentary "Fahrenheit 9/11", famously drove around Washington DC blasting the Patriot Act through a loudspeaker in order to show people what was actually in it (because it was just that damn ridiculous). Through a series of his interviews in the movie, it was even suggested that no Senator had even read the bill prior to voting on it. A Congressman stated, "We don't read most of the bills. Do you really know what that would entail if we read every bill that we passed? ... It would slow down the legislative process". God Bless America. And what's even funnier? The UK intelligence agencies are following suit, asking for relaxed regulations on surveillance laws, as they deem them "burdensome" when dealing with data gathering [16].

Now, while the Patriot Act was a (technically speaking) legal way of gathering private data and spying on people, there are also a ton of malware-based spying attempts (like those that Snowden revealed), such as Zeus, Pegasus, and Stuxnet [17–23]. Some of these have been proven to be made by governments themselves, others were made by individuals, but it was suspected to be used by governments. Let's unpack each of them quickly. Let's start with Zeus, a Trojan horse-type malware (a piece of software that looks harmless, but ends up doing something else after being downloaded), it spread over things like spam emails, pretending to be online stores, phone companies, banks, etc. Once a user clicked a link from one of these emails, they would be redirected to a link that would download and execute the program. Zeus would then monitor keystrokes (ideal for username and password theft), banking information, and all kinds of other goodies. It wasn't long before it ended up being used by governments searching for secret and sensitive information on compromised computers [17]. Searches could be specifically tailored to targeted countries, for instance, in Georgia (the country, not the state) they could look for information on specific government officials, Turkey could gather information on Syria, and everyone could search affected US computers for documents containing keywords like "top secret" and "Department of Defense".

Pegasus, on the other hand, was directly created for government use. Made by the Israeli company NSO, whose M.O. is aiding governments in combating terror and crime (emphasis on the dealing with government clients ONLY), the purpose of Pegasus is to be covertly and remotely installed on mobile phones running iOS and Android to read text messages, record calls, collect passwords, track the user's location, access the device's microphone and camera, and generally gather information from the installed apps [18, 19]. The funny thing is, even though NSO claims to make Pegasus as a product to fight crime and terrorism, it's been clearly evident that governments around the world have been using the program to do some pretty sketchy things, such as spy on journalists, lawyers, political dissidents, and even human rights activists [20]. To make matters even more interesting (or should we say, to muddy the waters even more)... although NSO is a private company, they cannot sell any licenses of Pegasus to foreign governments without approval from the Israeli defense ministry [21].

The last one of these spying softwares we will mention is Stuxnet. Stuxnet is a worm (essentially a virus that crawls from computer to computer on the same network) *ALLEGEDLY* built by a collaborative effort between the US and Israel, known as Operation Olympic Games, although neither have openly admitted this [22]. Essentially, Stuxnet targets supervisory control and data acquisition (SCADA) systems (a bunch of computers tied together for supervising and controlling some serious machines and processes, think factories or power plants). It has even done things like wreck Iran's nuclear facilities, where Stuxnet reportedly infected over 200,000 computers [23]. There was also an effort to do the same in North Korea, but it ended up failing due to their insane secrecy and isolation (it just wasn't possible to inject Stuxnet into their systems).

But, when it comes to stealing data from the government (or even a company or an individual), it actually isn't as difficult as you would think. You don't need these crazy global hacking attempts to access someone's data, as was shown in the 2015 show Mr. Robot. The main character, Elliott, needs to access data on a closed and secured police station network. So, to do this, he plays to his simplest advantage, the average human's curiosity (and stupidity). He uploads a virus onto a handful of USB drives and goes to the police station. He then proceeds to scatter them on the floor near the entrance, with the hopes that someone will pick one up and take it inside. Sure enough, one cop, bright-eyed and bushy-tailed, grabs one off the floor, takes it inside, and plugs it into his computer to see what's on it (as they say, curiosity killed the cat). At this point, the malware takes hold, and Elliott has access to the police station's network. And this isn't just a work of fiction. This, actually, is a very well-known tactic in real life. In 2008, a flash drive containing a worm called "Agent.btz" was plugged into a laptop at a US military base, and from there it spread like wildfire through the US Department of Defense's computers [24]. Once the compromise was detected, it took a glacial fourteen months to remove the malware from the government's networks. This, in fact, led to the ban of USB drives in many governmental offices and private companies. So, what did you learn from this? If you find some random USB drive, please, for the love of God, don't go plugging it into your system.

And now, lastly, let's talk about the scariest of all of these spying methods. The dreaded AI. And don't worry, you will learn much more about the potential terrors AI from Heba's chapter. So, let's just pick it up from ground zero. Now, we are surrounded by cameras… security cameras, traffic cameras, satellites, drones, phones, webcams, etc., both private and governmental. We know this, and this is no surprise. There's mass surveillance all over the world, be it in Shanghai, Abu Dhabi, or London. As of 2023, China has over 700 million surveillance cameras, and the US has about a quarter of that [25]. Generally, the laws surrounding the use of these types of recordings are fairly clear. What's done in public (generally speaking) can be recorded, what's done in areas that should be private, shouldn't (we'll get into this in a later section). As long as the use of CCTV doesn't really break these rules, and its use can be justified (such as to curb crime), its fine. But, nowadays, we're getting into a little bit of an ethical gray zone with the combination of these cameras, along with our good friend Artificial Intelligence.

With products such as Clearview AI, governments are able to use cameras to essentially predict criminal intent of behavior [26]. In addition, the company's algorithm can match real-time faces on cameras to a database of more than 20 billion images collected from the internet, including social media applications in order to identify criminals. The Ukrainian Ministry of Defense began using Clearview AI in 2022 to uncover Russian assailants and identify dead soldiers and prisoners of war [27]. Israel is also said to be using a similar AI-based surveillance technology named "The Gospel". The Israeli military claims to be able to cherry-pick their targets in real-time, identifying enemy combatants and equipment, all while reducing civilian casualties. However, many are arguing that this type of targeting system is "unproven at best – and at worst, providing a technological justification for the killing of… civilians" [28, 29]. The issue with this type of production-line killing technique is of course the rapid rate of the targeting process. While in the past, through traditional warfare, a military may find a new target every few days, with AI, this has jumped to hundreds of targets a day, and of course, many of these are unverified by the human eye. The system relies entirely on its programming, and has very little (to no) human intervention at all. As such, these systems can easily be viewed as an unbiased (or arguably a very biased) killing machine. That's the problem here, every new technological advantage gained can easily also become a new ethical or moral quandary.

But these types of AI-based systems don't have to be relegated to just war time. Governments all over the world are using them in a very George Orwell "1984"- style manner. For those of you that don't remember the basic premise of "1984", let me quickly remind you. In George Orwell's 1949 book, "1984", we (the collective we) live in a futuristic dystopian society. Civilization has been brought to tatters through a series of world wars. Three totalitarian super-states rule the world, and ours is called Oceania, led by "The Party" and the mysterious figurehead "Big Brother". The Party viciously eradicates anyone who does not conform to The Party's ideals, and uses constant surveillance (through two-way televisions, cameras, and hidden microphones) to spy on the population and flush out the infidels. The public is constantly reminded that they are being watched (as a warning to behave), through ubiquitous propaganda. "Big Brother is watching you". Those who fail to match the ideological purity of The Party become "unpersons", and disappear without a trace, with all evidence of their existence destroyed.

Now, in this book, the main issue was that there weren't enough people to review all of the data being gathered all of the time. And that's where AI comes in. AI can literally "watch" all of the surveillance cameras, review all of the data, and even make decisions based on it, with zero lunch breaks. And that's just what the likes of China, the US, and the UK are doing. Here is a quote directly from an NPR interview with the CEO of an AI surveillance company called Synthetaic [30]:

> For decades, cameras have been watching over cities, businesses and even homes. But that footage has mainly been stored locally, and reviewing it took a pair of human eyes. Not anymore. AI systems can now hunt for a van in a city, scan license plates and even faces in real time… Synthetaic really can find anything you want anywhere in the world… We've run searches, as an example, across the entire eastern seaboard of Russia for ships, and we can find every ship in a few minutes. It's pretty remarkable. Being able to scan the vast coastline of a nation like Russia is why this kind of technology has caught the eye of big government intelligence agencies. Watching everything that needs to be watched has always been a labor-intensive business… In

the digital age, intelligence agencies are drowning in photos and videos. There's just so much more data being collected than can realistically be analyzed by a human analysis workforce.

Now it's just a matter of time before it's implemented in a "1984"-style manner. In fact, there have been whispers about China implementing this type of monitoring in their social credit system (the one that gives their citizens a numerical score by analyzing their social behaviors in addition to looking at their fiscal and government data, and then accordingly punishing or rewarding them based on their score) [31, 32]. Imagine that, you gulp down that delicious can of Mountain Dew, and in a moment of weakness, you decide to toss it aside on the street, instead of throwing it away in a garbage bin like a decent human being. Immediately the CCTV picks this up, the AI behind it analyzes your face and figures out who you are, and BOOM! There goes your nice social credit score.

Now, realistically, these rumors have all been kind of dispersed, and it's been stated that China's social credit system is fairly low-tech [33]. The Chinese government also said that they are mainly interested in fraudulent and unethical business practices (like not honoring legal contracts, or defaulting on loans), and not really individuals that are littering their Mountain Dew cans on the mean streets of Wuhan. Obviously, once again, it's undeniable that there are benefits to these type of surveillance-based monitoring systems. But on the flip side, these systems can also be derailed and taken advantage of fairly easily. In cybersecurity, we always say, increased safety is a tradeoff with decreased privacy. So, you, dear reader, need to figure out which means more to you.

Now, as you can probably tell, the majority of this chapter (so far) has talked about how governments can spy on you by sneakily hijacking your internet communications and personal data. Of course, there are other types of spying as well, such as a workplace spying (where an employee is monitored by the employer), as well as just regular-old spying, where some rando is either being a Peeping Tom, or trying to gain access to your data through your devices, to even corporate-level spying, where companies want your data for one reason or another. Let's end this chapter by talking about each of these briefly.

EMPLOYEE MONITORING

The easiest of these to discuss is employee monitoring. It has been long since established that employers are able to monitor their employees while at work (to certain extent). Essentially, if you are on an employer-provided computer or phone, on an employer-furnished internet connection, the employer has every legal right to track what it is that you are doing (while on company time). For instance, this can be as simple as time tracking tools used by IBM [34], all the way to Amazon's new overlord "Mentor" which uses AI cameras to analyze and discipline their delivery drivers [35]. This practice can be traced all the way back to the early 1900s, when the use of punch cards became a thing. Employees had to clock in and clock out, but everything in between was still a mystery. You could literally clock in, fall asleep in a broom closet, and wake up and clock out at the end of the day. So, employers began thinking of alternate ways to track employee performance. Companies counted downtime in between phone calls for salespeople, words typed for office workers, production counts for factory employees, and so forth. All of this gave a rough estimate of the productivity of an individual, and their relative worth to the company.

But then came the internet, and with it a billion ways to waste time during work. Not only were employees less productive, they're also able to drag down the productivity of others by hogging bandwidth. As technology became more sophisticated, so did employee monitoring. Today, companies can use a ton of tools and techniques to monitor their employees, including time tracking, task monitoring, location tracking, video surveillance, and much, much more. Now, this all remains legal and above-bard, as long as the employer collects information about only work-related activities. But, sometimes, employers can (either inadvertently, or not) also collect employee's personal information. This becomes a balancing act of the two parties... Employees want to maintain their privacy, and employers want to make sure their company resources aren't misused. So, the rule of thumb here is, as long as the employee knows what is being monitored and how, and they agree to it, there is really no issues. If they don't like it, well, then they can always find a job elsewhere.

INDIVIDUALS SPYING ON OTHERS

Lady Godiva. Kind of like the chocolates, but the person. Raise your hand if you know where I'm going with this... To preface, Lady Godiva was an Anglo-Saxon noblewoman who was married to this dude named Leofric, who was the Earl of Mercia. So, as the story goes, Leofric liked his tax revenues. He was often portrayed as this unfeeling overlord who imposed over-taxation on his people. His wife, on the other hand, took pity on the people of Coventry, who were suffering under the oppressive taxation. Lady Godiva pleaded to her husband again and again, but Leofric simply refused to lower the taxes. Finally, after a bunch of her begging and pleading, Leofric said he would do it. But only if she rode through the streets of the town naked on a horse. Fair trade? So, Lady Godiva did just that but ordered everyone to shut their windows and stay indoors. Obviously, as the townsfolk wanted the tax relief, they did just that. All but this one guy. A tailor named Tom. He decided to take a look-see at what all this fuss was about. He took one peep (see what I did there). He became known as "Peeping Tom".

Nowadays, we have a slew of Peeping Toms all over the inter-net. There are some that are interested in just peeping at your data, but we've covered those in the identity theft chapter already, so let's glance over them. Then, there are those that want to peep at you. For instance, ex-Google employee David Barksdale, who snooped and stalked four underage users through his company access to their chats, emails, and other communications [36]. There have also been those that put up cameras in places where they shouldn't be, such as in bathrooms, changing rooms, or even hotel rooms (a whole heap of these have been surfacing with Airbnb's popularity on the rise) [37]. Basically, people install cameras or microphones in private spaces to view and record what goes on. What they choose to do with this footage, we'll talk about shortly. Then, there are the cameras that YOU already have. These can be baby monitors, security cameras, those fancy Nest doorbells, and of course, your webcams (and yes, that includes all four cameras on your iPhone 14 Pro Max). People are doing all of these for three main reasons, either just to get their jollies off, blackmail, or sometimes, something even more sinister.

"Wake up little boy, daddy's looking for you". This was the sentence that came through a 3-year-old's baby monitor in Washington [38]. In Texas, a mother heard someone calling her 2-year-old daughter a "moron". And in Indiana, parents heard someone playing that one song by The Police that, in this given situation, was very fitting... "Every move you make, Every step you take, I'll be watching you". These attacks are called camfecting. But it gets worse than that, security camera footage can be hacked into to see if anyone is home, in order to break into a house. Baby monitor footage can even be used to plan a child abduction [39]. Or, like the case with 2013 Miss Teen USA, Cassidy Wolf, webcams can be hacked to simply watch you [40]. In Cassidy's situation, one of her classmates silently watched her through her webcam in her bedroom for days. He even went so far as to emailing the photos to her and threatening to release them to the public, if she didn't undress for him in front of the camera. She ended up filing a complaint with the

FBI, who prosecuted the hacker, and he was sentenced to 18 months in prison. But what's to stop him from doing it again to someone else? And what's the guarantee that the person being spied on even knows that someone is peeping in on them?

COMPANIES SPYING ON INDIVIDUALS

The last of these non-governmental spying stories is something that is unfolding right now, as you read this. When you buy that new iPad Air for that massive $599 price tag, and unwrap the beautiful cardboard box, and press that power button for the very first time, you are greeted with a nice, lengthy contract for all of Apple's Terms and Conditions. Do you read it? Who does? You do just like the Congresspeople did when they voted on the Patriot Act, scroll right on through and accept. Well, what is lurking in the shadows? Well, it's the deepest, darkest caveats!

Enable Siri? Yes, please! Ask Alexa? Why not?! All of your devices now have microphones built in, so that you can use these handy dandy virtual assistants. At least that's what the companies tell you. But, had you read that contract a little bit carefully, you would have realized that you have given Apple [41]... or it could be Amazon [42], or Google [43], or whoever, the permission to unbeknownst to you, record your voice. Want your boyfriend to propose to you? Grab his phone and say "engagement rings" into it ten times. Next time he goes and visits a website with paid ads, guess what he's going to see? Yep, a link to buy that shiny new platinum Tiffany setting with that three-carat whopper on top. But it goes beyond this. It's not just your voice that companies are recording, but where you shop, what you order, where you eat, what emails you're sending (lookin' at your Gmail), what websites you visit, what posts you double-tapped on your Insta, and so forth. From this, companies are able to build a very nice profile of you, through which they can target you with the exact products you are most likely to trade your hard-earned cash for [44, 45].

Speaking of Target (now we're onto the American retail store, not the darts bullseye), the last anecdote of this chapter is a doozy. The New York Times published a story a while back called "How Companies Learn Your Secrets" [44]. Through some maths and analytics, Target came to the conclusion that new parents are a retailer's holy grail. Most shoppers don't buy everything in one store, and this is generally an ingrained habit that is incredibly difficult to change. There are, however, some brief moments in time in a person's life when these old habits fall apart. Target found out that one of those moments is when someone finds out they're about to have a baby. In this golden moment, anything goes, and brand loyalty is up for grabs. The problem is, timing is key. As soon as a birth is public, everyone knows, and the new mother is bombarded with ads. So, Target needs to reach them before. Perhaps before even the mother knows she's pregnant. Is that even possible, you may ask? Well, according to Target it is. Through a series of data points, like what you buy, when you buy it, where you live, how old you are, the credit card you use, etc. [45]:

> Take a fictional Target shopper named Jenny Ward, who is 23, lives in Atlanta and in March bought cocoa-butter lotion, a purse large enough to double as a diaper bag, zinc and magnesium supplements and a bright blue rug. There's, say, an 87 percent chance that she's pregnant and that her delivery date is sometime in late August.

Target built a model to fairly accurately predict when a soon-to-be mother was in her second trimester of pregnancy. Not only that, they had a general idea of when the baby was due as well. Now, the issue was, they couldn't just outright yell at people "Hey, congrats on your pregnancy, ma'am!", because that would be a PR nightmare. It would appear that Target was spying on their customers (which of course, to a certain level they were). As an executive for Target said [44]:

With the pregnancy products, though, we learned that some women react badly. Then we started mixing in all these ads for things we knew pregnant women would never buy, so the baby ads looked random. We'd put an ad for a lawn mower next to diapers. We'd put a coupon for wineglasses next to infant clothes. That way, it looked like all the products were chosen by chance. And we found out that as long as a pregnant woman thinks she hasn't been spied on, she'll use the coupons. She just assumes that everyone else on her block got the same mailer for diapers and cribs. As long as we don't spook her, it works.

Funny isn't it, that this kind of human behavior tracking is perfectly legal, and technically NOT considered "spying" (by the dictionary definition, "spying" is just collecting information about something to use in deciding how to act, and Target hits this definition spot on). But if customers knew what was going on, even Target admits, they wouldn't be happy.

So, this brings me to the story... While Target was still fine-tuning this algorithm, before they found out that they shouldn't just send ads and coupons for baby products to customers, they, of course, did try just that. Well, a man shows up to a Minneapolis Target super pissed off, demanding to see the manager and holding a stack of coupons. "My daughter got this in the mail! She's still in high school, and you're sending her coupons for baby clothes and cribs? Are you trying to encourage her to get pregnant?" Of course, the manager had no idea what the man was on about. He wasn't in the loop on the new-fangled advertising tactics Target was using. But, it was a Target advertisement, sent to the man's daughter with a bunch of coupons for maternity clothing, nursery furniture, and pictures of happy, smiley babies. All the manager could do was apologize. He had no idea what was going on... A few days later the manager called the man to apologize again. Just to be on the safe side, and make sure there were no hard feelings. The man sounded off and somewhat embarrassed. "I had a talk with my daughter", he said. "It turns out there's been some activities in my house I haven't been completely aware of. She's due in August" [44, 45].

Now, at the end of all this, what's the takeaway? Is there a way of getting around all of these types of spying? Sure, you can use antivirus programs, and anti-spywares, tape up your webcams, stop using baby monitors, shit, you can move into an igloo in Alaska. But at the end of the day, once you step foot outside the boundaries of your igloo, it's free range. These days, anything that is public is monitored and recorded for quality assurance purposes. Anything that is private, is hell, probably still public, and monitored just to be on the safe side. I always say, send every email as if it's on company letterhead. Make every phone call as if it's being recorded. Whatever data someone has about you, just make sure you give them the least possible ammunition against you. In the end, there's no use in fighting it. Relax. Accept the inevitable. And remember: Big Brother is watching you.

REFERENCES

[1] "Edward Snowden," Wikipedia. [Online]. Available: https://en.wikipedia.org/wiki/Edward_Snowden. [Accessed: Jun. 19, 2024].

[2] S. D. Dwilson, "Was Snowden really injured in the army?" Heavy, Sep. 16, 2016. [Online]. Available: https://heavy.com/entertainment/2016/09/was-snowden-injured-in-army-did-break-legs-how-shin-splints-special-forces-movie/. [Accessed: Jun. 19, 2024].

[3] "Edward Snowden: A timeline," NBC News, May 27, 2014. [Online]. Available: https://www.nbcnews.com/feature/edward-snowden-interview/edward-snowden-timeline-n114871. [Accessed: Jun. 19, 2024].

[4] G. Greenwald, "NSA collecting phone records of millions of Verizon customers daily," The Guardian, Jun. 6, 2013. [Online]. Available: https://www.theguardian.com/world/2013/jun/06/nsa-phone-records-verizon-court-order. [Accessed: Jun. 19, 2024].

[5] "Booz Allen Statement on Reports of Leaked Information," Booz Allen Hamilton, Jun. 11, 2013. [Online]. Available: https://investors.boozallen.com/news-releases/news-release-details/booz-allen-statement-reports-leaked-information. [Accessed: Jun. 19, 2024].

[6] N. Hopkins and J. Borger, "Exclusive: NSA pays £100m in secret funding for GCHQ," The Guardian, Aug. 1, 2013. [Online]. Available: https://www.theguardian.com/uk-news/2013/aug/01/nsa-paid-gchq-spying-edward-snowden. [Accessed: Jun. 19, 2024].

[7] B. Gellman and A. Soltani, "NSA collects millions of e-mail address books globally," Washington Post, Oct. 14, 2013. [Online]. Available: https://www.washingtonpost.com/world/national-security/nsa-collects-millions-of-e-mail-address-books-globally/2013/10/14/8e58b5be-34f9-11e3-80c6-7e6dd8d22d8f_story.html. [Accessed: Jun. 19, 2024].

[8] A. Anishchuck, "Snowden receives three-year Russian residence permit - lawyer," Reuters, Aug. 7, 2014. [Online]. Available: https://www.reuters.com/article/idUSKBN0G70ZR/. [Accessed: Jun. 19, 2024].

[9] M. Ilyushina, "Edward Snowden gets permanent residency in Russia - lawyer," CNN, Oct. 22, 2020. [Online]. Available: https://edition.cnn.com/2020/10/22/europe/edward-snowden-russia-residency-intl/index.html. [Accessed: Jun. 19, 2024].

[10] "Putin grants Russian citizenship to U.S. whistleblower Snowden," Reuters, Sep. 27, 2022. [Online]. Available: https://www.reuters.com/world/europe/putin-grants-russian-citizenship-us-whistleblower-edward-snowden-2022-09-26/. [Accessed: Jun. 19, 2024].

[11] "Why Wikileaks' Julian Assange faces US extradition demand," BBC News, May 20, 2024. [Online]. Available: https://www.bbc.com/news/world-us-canada-68282613#. [Accessed: Jun. 19, 2024].

[12] "Constitutional Amendments – Amendment 4 – 'The Right to Privacy'," Ronald Reagan Presidential Library. [Online]. Available: https://www.reaganlibrary.gov/constitutional-amendments-amendment-4-right-privacy. [Accessed: Jun. 19, 2024].

[13] M. Tran, "Bush signs controversial surveillance bill," The Guardian, Aug. 6, 2007. [Online]. Available: https://www.theguardian.com/world/2007/aug/06/usa.marktran1. [Accessed: Jun. 19, 2024].

[14] B. R. Lemons, "The U.S. Patriot Act of 2001 changes to electronic surveillance laws," Federal Law Enforcement Training Centers, 2001. [Online]. Available: https://www.fletc.gov/sites/default/files/imported_files/training/programs/legal-division/downloads-articles-and-faqs/research-by-subject/miscellaneous/patriotact.pdf. [Accessed: Jun. 19, 2024].

[15] D. Lithwick and J. Turner, "A guide to the Patriot Act, Part 1," Slate, Sep. 8, 2003. [Online]. Available: https://slate.com/news-and-politics/2003/09/a-guide-to-the-patriot-act-part-1.html. [Accessed: Jun. 19, 2024].

[16] H. Davies, "UK spy agencies 'broke privacy rules' on AI data," The Guardian, Aug. 1, 2023. [Online]. Available: https://www.theguardian.com/technology/2023/aug/01/uk-intelligence-spy-agencies-relax-burdensome-laws-ai-data-bpds. [Accessed: Jun. 19, 2024].

[17] M. Schwirtz, "Russian Espionage Piggybacks on a Cybercriminal's Hacking," New York Times, Mar. 12, 2017. [Online]. Available: https://www.nytimes.com/2017/03/12/world/europe/russia-hacker-evgeniy-bogachev.html?smid=tw-nytimes&smtyp=cur&mtrref=undefined&_r=3. [Accessed: Jun. 19, 2024].

[18] T. Brewster, "Everything we know about NSO group, the professional spies who hacked iPhones with a single text," Forbes, Aug. 30, 2016. [Online]. Available: https://www.forbes.com/sites/thomasbrewster/2016/08/25/everything-we-know-about-nso-group-the-professional-spies-who-hacked-iphones-with-a-single-text/?sh=741751df3997. [Accessed: Jun. 19, 2024].

[19] C. Timberg, R. Albergotti, and E. Gueguen, "Despite the hype, iPhone security no match for NSO spyware," Washington Post, Jul. 19, 2021. [Online]. Available: https://www.washingtonpost.com/technology/2021/07/19/apple-iphone-nso/. [Accessed: Jun. 19, 2024].

[20] R. Bergman and M. Mazzetti, "The battle for the world's most powerful cyberweapon," New York Times, Jun. 15, 2023. [Online]. Available: https://www.nytimes.com/2022/01/28/magazine/nso-group-israel-spyware.html. [Accessed: Jun. 19, 2024]

[21] D. E. Sanger, N. Perlroth, A. Swanson, and R. Bergman, "U.S. Blacklists Israeli Firm NSO Group Over Spyware," New York Times, Nov. 3, 2021. [Online]. Available: https://

www.nytimes.com/2021/11/03/business/nso-group-spyware-blacklist.html. [Accessed: Jun. 19, 2024].

[22] E. Nakashima and J. Warrick, "Stuxnet was work of U.S. and Israeli experts, officials say," Washington Post, Jun. 2, 2012. [Online]. Available: https://www.washingtonpost.com/world/national-security/stuxnet-was-work-of-us-and-israeli-experts-officials-say/2012/06/01/gJQAlnEy6U_story.html. [Accessed: Jun. 19, 2024].

[23] M. B. Kelley, "The Stuxnet attack on Iran's Nuclear Plant Was 'Far More Dangerous' than previously thought," Business Insider, Nov. 20, 2013. [Online]. Available: https://www.businessinsider.com/stuxnet-was-far-more-dangerous-than-previous-thought-2013-11. [Accessed: Jun. 19, 2024].

[24] N. Shachtman, "Insiders doubt 2008 pentagon hack was foreign spy attack (Updated)," Wired, Aug. 25, 2010. [Online]. Available: https://www.wired.com/2010/08/insiders-doubt-2008-pentagon-hack-was-foreign-spy-attack/. [Accessed: Jun. 19, 2024].

[25] "Mass surveillance in China," Wikipedia. [Online]. Available: https://en.wikipedia.org/wiki/Mass_surveillance_in_China. [Accessed: Jun. 19, 2024].

[26] J. R. Saura, D. Ribeiro-Soriano, and D. Palacios-Marqués, "Assessing behavioral data science privacy issues in government artificial intelligence deployment," *Government Information Quarterly*, vol. 39, no. 4, p. 101679, 2022.

[27] K. Hill, "Facial recognition goes to war," New York Times, Apr. 7, 2022. [Online]. Available: https://www.nytimes.com/2022/04/07/technology/facial-recognition-ukraine-clearview.html. [Accessed: Jun. 19, 2024].

[28] G. Brumfiel, "Israel is using an AI system to find targets in Gaza. Experts say it's just the start," NPR, Dec. 14, 2023. [Online]. Available: https://www.npr.org/2023/12/14/1218643254/israel-is-using-an-ai-system-to-find-targets-in-gaza-experts-say-its-just-the-st. [Accessed: Jun. 19, 2024].

[29] H. Davies, B. McKernan, and D. Sabbagh, "'The Gospel': How Israel uses AI to select bombing targets," The Guardian, Dec. 1, 2023. [Online]. Available: https://www.theguardian.com/world/2023/dec/01/the-gospel-how-israel-uses-ai-to-select-bombing-targets. [Accessed: Jun. 19, 2024].

[30] G. Brumfiel, "How AI is revolutionizing how governments conduct surveillance," NPR, Jun. 13, 2023. [Online]. Available: https://www.npr.org/2023/06/13/1181868277/how-ai-is-revolutionizing-how-governments-conduct-surveillance. [Accessed: Jun. 19, 2024].

[31] N. Kobie, "The complicated truth about China's social credit system," Wired, Jun. 7, 2019. [Online]. Available: https://www.wired.com/story/china-social-credit-system-explained/. [Accessed: Jun. 19, 2024].

[32] "CHINA'S SOCIAL CREDIT SYSTEM," Bertelsmann Stiftung. [Online]. Available: https://www.bertelsmann-stiftung.de/fileadmin/files/imported_files/aam/Asia-Book_A_03_China_Social_Credit_System.pdf. [Accessed: Jun. 19, 2024].

[33] "China's Social Credit Score: Untangling Myth from Reality," Mercator Institute for China Studies, Feb. 11, 2022. [Online]. Available: https://merics.org/en/comment/chinas-social-credit-score-untangling-myth-reality. [Accessed: Jun. 19, 2024].

[34] "What is employee productivity?," IBM, [Online]. Available: https://www.ibm.com/topics/employee-productivity. [Accessed: Jun. 19, 2024].

[35] A. Palmer, "Amazon uses an app called Mentor to track and discipline delivery drivers," CNBC, Feb. 12, 2021. [Online]. Available: https://www.cnbc.com/2021/02/12/amazon-mentor-app-tracks-and-disciplines-delivery-drivers.html. [Accessed: Jun. 19, 2024].

[36] K. Zetter, "Ex-Googler allegedly spied on user E-Mails, chats," Wired, Sep. 15, 2010. [Online]. Available: https://www.wired.com/2010/09/google-spy/. [Accessed: Jun. 19, 2024].

[37] H. Fry, "This is where some Airbnb hosts have been spying on you, and why it will stop," Los Angeles Times, Mar. 13, 2024. [Online]. Available: https://www.latimes.com/california/story/2024-03-13/feel-like-somebodys-watching-you-inside-an-airbnb-not-for-long-company-says. [Accessed: Jun. 19, 2024].

[38] J. Flannigan, "Parental warning: Your baby monitor can be hacked," HuffPost, Aug. 24, 2017. [Online]. Available: https://www.huffpost.com/entry/parental-warning-your-bab_b_11668882. [Accessed: Jun. 19, 2024].

[39] K. Utehs, "Baby cam monitor hack and kidnapping threat serve as warning for tech security," ABC7 News, Dec. 21, 2018. [Online]. Available: https://abc7news.com/baby-monitor-cam-hack-threat/4940379/. [Accessed: Jun. 19, 2024].

[40] S. Nelson, "Tape Your Webcam? 'Horrifying' Malware Broadcasts You to the World," US News & World Report, Jul. 29, 2015. [Online]. Available: https://www.usnews.com/news/articles/2015/07/29/tape-your-webcam-horrifying-malware-broadcasts-you-to-the-world. [Accessed: Jun. 19, 2024].

[41] K. O'Flaherty, "Apple siri eavesdropping puts millions of users at risk," Forbes, Jul. 28, 2019. [Online]. Available: https://www.forbes.com/sites/kateoflahertyuk/2019/07/28/apple-siri-eavesdropping-puts-millions-of-users-at-risk/. [Accessed: Jun. 19, 2024].

[42] G. A. Fowler, "Alexa has been eavesdropping on you this whole time," Washington Post, May 6, 2019. [Online]. Available: https://www.washingtonpost.com/technology/2019/05/06/alexa-has-been-eavesdropping-you-this-whole-time/. [Accessed: Jun. 19, 2024].

[43] D. MacMillan, "Tech's dirty secret: The App developers sifting through your Gmail," Wall Street Journal, Jul. 2, 2018. [Online]. Available: https://www.wsj.com/articles/techs-dirty-secret-the-app-developers-sifting-through-your-gmail-1530544442. [Accessed: Jun. 19, 2024].

[44] C. Duhigg, "How companies learn your secrets," New York Times, Feb. 16, 2012. [Online]. Available: https://www.nytimes.com/2012/02/19/magazine/shopping-habits.html. [Accessed: Jun. 19, 2024].

[45] K. Hill, "How target figured out a teen girl was pregnant before her father did," Forbes, Aug. 11, 2022. [Online]. Available: https://www.forbes.com/sites/kashmirhill/2012/02/16/how-target-figured-out-a-teen-girl-was-pregnant-before-her-father-did/?sh=6791c59c6668. [Accessed: Jun. 19, 2024].

7 Bytes and Blues – How the Internet Plays with Adult Minds

Picture a small village in rural Ireland in the 1970s. One school, one church (Catholic of course), one television station that broadcast what seemed like a lot of farming news but there were 13 bars for a population of just several hundred people. There wasn't really much else to do. That began to change a little when in 1978 a second TV station launched that was quite a bit different from the one that already existed. This one had a bit more of a focus on the external world and introduced us, inhabitants of the village, to the larger world in general. It also introduced us to the very specific world of the 'Moonies' or to give them their official title, members of the Unification Church of South Korea. The church was founded by the Reverand Sun Myung Moon in the 1950s, which is why the thousands of people who joined the church in the 1970s came to be known by the term 'Moonies'. Irish TV certainly wasn't promoting the Unification Church, but it just happened at one point to televise a totally new phenomenon to us; the mass wedding. In Madison Square Garden, New York in 1982, footage was televised of 2075 couples (that's 4150 people) participating in a single mass wedding.

What was striking at the time was the way everyone was dressed. All the women seemed to be wearing the same wedding dress and all had a white fluffy tiara across their heads. All the

DOI: 10.1201/9781032679389-8

men wore black suits and for some reason white gloves. The only questions the other kids my age were asking each other were why did everyone look the same and why were thousands of people getting married at the same time. Irish weddings were always single-couple events. But the internet (and a decent library) didn't exist back in 1982 and it would be years later before it was possible to dig a little deeper. Today, the Unification Church of South Korea is considered by most, except by those people in it of course, to be a cult. Cults are truly fascinating things to study and membership of one can sometimes result in people doing seemingly bizarre and strange things, like participating in a mass marriage. But sometimes, the results can be a lot more destructive and even lethal. The most infamous example of this is the mass suicide that took place in Guyana, South America in 1978. In the previous year, several hundred members of an American cult called the 'Peoples Temple' established a settlement in the jungle under the leadership of a man called Jim Jones who, of course, promised it would be a utopia for everyone. Except, it wasn't and the following year, Jones convinced many of the members of the Peoples Temple to commit 'revolutionary suicide'. Many did and those who didn't were murdered by those who were more committed to the cause. In all 918 people died mostly by poisoning, including 300 under the age of 17 [1]. Several things connect the Unification Church and the Peoples Temple but perhaps the single most important one is that both leaders, Moon and Jones, were in fact very effective influencers. People believed them and they followed them. Moon claimed to possess some serious heavy-hitting credentials, including being a messiah, that he was immaculate and incapable of sin. According to Moon, Jesus recognized him as the savior of humankind, and so had the Buddha, Satan and others [2]. Jones also possessed some equally impressive credentials including also being the messiah, a performer of miracles and the only person with the knowledge to survive the coming apocalypse [3].

ONLINE CULTS

You may well ask about the relevance of cults from the 1970s and 1980s as that's just history. However, fast forward to August of 2023 when a group of six people including two children go missing in Missouri, USA. All were living in a house in St. Louis when one day they all disappeared. Food was left in the microwave, laundry was still in the washing machine and somewhat strangely, shoes worth several thousand dollars were left behind [4]. Everyone had clearly left in a hurry. A subsequent police investigation claimed that the missing group had previously become part of an online spiritual cult following the teachings of a man called Rashad Jamal. Similar to people like Moon and Jones, Jamal also claims to have some impressive credentials including being a semi-divine human being sent back to Earth. He has stated that his task is to 'enlighten and inform and increase the frequency of the planet' and to elucidate people with the truth about the government which apparently is deliberately shutting off stargates to other dimensions.

However, there is one big difference between cults from the 1970s and the present day. Whereas the members and leaders of the Peoples Temple and the Unification Church congregated in a physical location, today people are not confined by physical limitations and can come together in the virtual world. Jamal's 'teachings' can be found on his YouTube channel called the University of Cosmic Intelligence which has an impressive sum of 207,000 subscribers and as of writing 204 'educational' videos available to view. He has nearly 200,000 followers on TikTok, more than 10,000 followers on X and 90,000 followers on Instagram [5]. By any standards, that's quite the following and quite the influence. The Unification Church or the Peoples Temple never had that level of following.

It's also a considerable achievement for Jamal who is currently serving 18 years in a federal prison in Georgia for child molestation and child cruelty. Somehow, he occasionally seems to continue to post his teachings online from inside the prison. Jamal says he's not a cult leader but merely a 'spiritual influencer' bringing the truth of the cosmos to his followers. One difference between cults or 'high-control groups' today and in the past is that these days you might never even meet the cult leader in person (like Jamal claims in respect of the six missing people from St. Louis). They can be virtually on a screen in your home with Hollywood-level video productions at any time courtesy of platforms like YouTube. That's a feature of the internet today, the sheer extent to which more 'unorthodox' but highly polished visualized ideas can spread and be embraced by those searching for meaning, answers and connection. After all, the number one reason adults use the internet is to find information [6].

Dismissing cults as some form of aberration or abnormality and classing their members as unhinged is a mistake because the study of cults can in fact can tell us a lot about our needs as human beings, and even how we continue to meet them in the age of the internet. The internet is new but the needs we have are very old and one of those fundamental needs we all have is a need for belonging. Why sit at home all sad and lonely when there's this group of lovely, friendly people just down the road promising all sorts of amazing benefits or now, just a click away online? Belonging to a group can also have major additional perks such as providing an understanding of what can often be a very confusing world, and even solutions to your problems in life. They often promise to provide you with knowledge about something bigger than yourself; the real machinations of the government, the truth of the universe, and the opportunity to find true love. For the latter, look no further than the 'Twin Flames Universe' which offers you the opportunity (for a substantial fee) to find 'your ultimate lover and eternal spiritual complement, your perfect person' [7]. As a bonus, you can also be part of 'creating a New Earth'. It all sounds wonderful, except former members of the organization allege that the two self-proclaimed gurus, and owners, of the 'Universe' use cult-like coercion to control every aspect of their followers' lives, ranging from regulating their level of contact with their families of origin, to selecting their romantic partners, and even so far as assigning gender identities based on little more than what seems like a whim [8]. It's also a masterclass example in how to hijack the internet for sophisticated marketing via YouTube, Instagram, X, Facebook and their own dedicated website. YouTube alone hosts 1,506 Twin Flames videos and not one appears to have even a hint of negativity about the group. They are likely to have been screened out, deleted or they don't appear because the threat of legal action is too intimidating. Little wonder that the owners of the Universe are now millionaires. Despite the presence of two damning documentaries about the group on Netflix and Amazon Prime, people continue to join.

Cults, whether they be in the real world or in the virtual world typically arise from a mix of several basic ingredients. First, we are primarily social beings and almost all of us have a desire to simply to be with someone or connect to something, whether it be a cause, a group or a shared set of ideas. Second, we are meaning machines; we need and seek explanations for the often-disparate events that surround us. Added to that, we are often drawn to people who seem to have the answers and are convincing enough to pull it off. Third, add into that mix feelings of being disenfranchised, disempowered, disconnected, deprived and dismissed and the space for being 'influenced' is wide open.

It's interesting to listen to people who have been in cults describing why they got into them in the first place. Many describe a feeling of being personally lost and disconnected, and what got them into the cult and kept them in it was a sense of connection with other cult members and the cult leaders themselves. They typically describe a strong sense of community, albeit an often dysfunctional one, characterized by high levels of coercive control. In essence, their reasons sound very much like a sense of belonging or meeting a social need. It's an often-unrecognized fact that we are more heavily influenced by others than we often care or wish to admit [9]. The social bonds that keep people in cults are strong, a sense of community, like-minded people and a sense of shared purpose. It doesn't really matter if it's a religious cult, an anti-government cult or even these days a political cult. They all meet this need to belong to a community, to be connected to others and today you can get it all online.

THE EARLY PROMISE OF CONNECTION

If you look at the early manifestations of social media on the internet like Six Degrees, Frendster, Facebook, MySpace, Hi5, all had a common denominator which was the promise of increased connection to others. They offered the means by which we could talk with each other online. They circumvented the difficulties imposed by physical distance and soon allowed us the opportunity to engage with old friends while at the same time offering the opportunity to connect with a community of others with similar interests whom we might have never otherwise met [10]. Back then, on those platforms, the social element was mostly about sharing photos of your latest holiday. In December of 2005, Facebook had 6 million active users. In 2024 that figure has risen to over 3 billion monthly users. Read between the lines: that's 3 billion people at least trying to meet a social need in some shape or form. In fact, globally, staying in touch with friends and family is in fact the second most cited reason people use the internet after finding information [11].

However, social media as a means of connecting people soon morphed into something quite different from what was originally intended. It quickly became a space for a lot more than simply connecting with others. It became a means of self-embellishment, 'alternative' news dissemination, selling products and activities, branding, and advertising. It later generated features such as the 'like', the 'follower' and of course the now

near-ubiquitous term 'influencer'. It also became, as already discussed, a place where people offered meaning and answers, and
more often than not for a price.

What only became apparent later with social media was that
once you open up the door for virtual social connection, you
also unleash a range of other normal human vulnerabilities into
the virtual or online world such as the need for social validation,
and the inevitable activities of social comparison and social
upmanship. The internet, and social media in particular, has provided a new playground for these and other vulnerabilities to
play themselves out. Three billion people might appear to be
connected on Facebook but the important question to ask is
whether this new connectedness is doing any good to people's
well-being. It's a question that has been the focus of a lot of
research over the past 20 years and the answer is a complex one.
But the question as to whether social media usage is bad for
people is perhaps better reframed as a question of how it interacts with normal human characteristics and vulnerabilities. For
example, these days we have new phrases to describe some of
the unforeseen effects of social media usage such as 'Facebook
Envy' and 'Facebook Depression'. How have these terms come
into everyday language?

EFFECTS OF SOCIAL COMPARISON

Well, one of the single most reliable and surefire ways to make
yourself feel unhappy is to compare yourself unfavorably with
others. Whereas in the past you might feel a little jealous over
your next-door neighbor's new car parked outside their door,
at least you were just comparing yourself to someone in close
physical proximity to you. However, social media platforms
can put this ordinary trait of social comparison on steroids
with often unhappy results. Now there is almost no limit to the
number of people you can compare yourself against in a virtual
world except there is one fundamental caveat. When scrolling

through the profiles of others (which is called Facebooking), what you are generally exposed to is only the positive aspects of other people's lives, the amazing activities they appear to engage in, the cool experiences they have, the seemingly perfect families and lives that just appear to be happier and better than yours. Because it's easy to create media quickly and easily on phones, tablets and home computers, what people began posting on social media was faster, shorter and a lot less authentic and people began posting and portraying a more aspirational view of themselves [12]. High levels of exposure to this kind of super positive but frequently very skewed highlights of the lives of others will more often than not result in feelings of inferiority. But the effect is amplified by the fact that most users of social media platforms like Facebook typically only post the positive aspects of their lives while excluding the negative. When you repeatedly feel inferior, depression will not be lurking far behind [13]. You might think that people would recognize what's happening to them and stop scrolling on social media, but many don't and keep going back often in a vain attempt to find evidence that other people's lives are not in fact better than their own. In response, some even turn to boasting about, or amplifying, their achievements to portray themselves in a better light which contributes to the vicious cycle of comparison. Curiously, this typically manifests itself somewhat differently across genders such as men tending to post more about their accomplishments while women tend to emphasize their appearance and social lives [14].

One early study on Facebook envy in Germany found that as many as one out of three people actually felt worse and less satisfied with their own lives after spending time on Facebook [15]. Believe it or not, the most common cause of feelings of resentment and jealousy, among Germans at least, was apparently seeing holiday photos of others. After that, indicators of having a better social life were the second biggest cause of envy, as users even compared seemingly innocuous things like the number of birthday greetings, likes, and comments made by others to their online friends. Not surprisingly, the social comparisons that people engage in typically reflect the general ambitions characteristic of their age group. For example, people in their mid-30s are most likely to envy family happiness while younger women are more likely to envy physical attractiveness of other women. Men in their 30s on the other hand tend to envy social status and prestige of other men on social media [16]. The key thing to remember though is that social comparison is an activity that we all engage in and in many ways, it's a perfectly normal thing. While social comparison can prompt some people to move their lives in a positive direction, it can also elicit among many others negative feelings of envy, jealousy and resentment.

However, the big change now is the context where social comparison is happening. It's an unwinnable game: An endless supply of online opportunities for comparison with a virtual but often warped version of reality. Given that the average adult spends approximately 143 minutes per day on social media it's easy to see how it might foster the creation of unrealistic expectations on what people think they should own, how they think they should look, or what they believe they need to achieve in life and by when [17]. As such, social media can operate as an incubator for unrealistic and often unattainable expectations which typically results in disappointment, frustration, anxiety and a constant sense of inadequacy which is not good for anyone's mental health. Unfortunately, it seems that some age groups appear to be more affected than others.

AGE AND MENTAL HEALTH

One unpleasant global trend in the 21st century is that the mental health of young adults (those aged between 18 and 25 years) seems to have fallen off a cliff. Compared to previous generations, higher and higher numbers of individuals in this stage of life are reporting more anxiety, loneliness, depression and stress but also less meaning in life [18]. While the causes of such problems are complex, some point the finger toward the effects of social media usage in this group who are the first generation to fully experience the digital world since birth. The childhood and adolescent stages of this age cohort have been characterized by a near-constant exposure to a steady diet of social media in a way never seen before and the effects of which are likely to have been carried into young adulthood. Despite the inherent promise of increased connection from social media, increasing numbers of young adults themselves identify social media as a contributory factor to impoverished mental health because it facilitates rampant social comparison [19]. They also often report reduced face-to-face interaction, increased alienation, negative self-perception, and problems with self-esteem and body image as a result of heavy social media usage.

There are big differences between the respective popularity of social media platforms across various stages of adulthood and depending on age, they seem to have different effects. Instagram, TikTok and Snapchat are the most popular among those under 24 whereas they barely register for those over 65 [20]. The trend reverses for Facebook and in the United Kingdom less than 10% of people under the age of 24 report it as their preferred platform. In contrast, over three-quarters of those aged over 65 report it as their preferred platform. Higher levels of social media usage among young adults are consistently associated with poorer mental health for a significant proportion of this group but increasing usage for those aged over 50 seems to have a positive effect. In fact, it seems to be that infrequent use of the internet is associated with deteriorating levels of life satisfaction among older people [21]. One reason why older users of social media generally seem to be mentally better off is simply because the older you get, the less likely you are to compare yourself against others and you're less likely to care what others think of you [22].

SOCIAL MEDIA AND RELATIONSHIPS

It's not just individuals that social media usage is having an impact on, it's also increasingly recognized that it is having some negative effects on relationships. If you are married or thinking about getting married (or even on the path to divorce), you might want to consider a rather unfortunate new trend which is that social media is increasingly a factor associated with an increasing number of divorces. A quick scroll through the websites of many legal firms points to at least anecdotal evidence of a rise in the number of clients who claim that Facebook, Snapchat, Twitter, WhatsApp or other social media networks had played a significant part in their divorce. While it might sound a bit sensational, some of the reported figures are, in fact, not trivial. Some sources suggest that up to one-third of divorces cite Facebook as a contributing factor to the breakup [23]. Nor is it just an American phenomenon as around 20% of divorce petitions included references to Facebook according to a 2009 survey in the UK [24]. Perhaps a little more worrying is that two-thirds of legal professionals say they use the social media profiles of one or another partner in divorce proceedings. It's easy to see why. A husband who pleads financial impoverishment in a courtroom can find himself having a somewhat difficult time explaining pictures of himself on social media showing his recent holiday in the Seychelles while standing in front of a judge. Instead of thinking before you jump, it might be wiser these days to think before you post. The good news is that legal firms now even offer advice on what to do and what not to do on social media if you are facing a divorce. For example, if you find yourself thinking about deleting your social media accounts in the midst of a divorce, then the legal advice is don't do it. That's because in countries like the USA, deletion of social media accounts constitutes 'spoliation' of evidence in divorce, which judges don't look too kindly on [25].

So do social media networks, like Facebook, have a detrimental effect on relationships and marriages? It's easy to come across research studies reporting that higher social media usage by one or both people in a romantic relationship results in higher levels of conflict in relationships, that partners feel neglected and there are lower levels of commitment and lower feelings of passion and intimacy in the relationship [26]. All of this might lead you to think that social media is somehow directly responsible. However, what is probably more accurate is that social networking is particularly dangerous for relationships that have already hit a rocky patch which they inevitably do at some point. If a couple is in that space, then normal vulnerabilities become activated and become focused on the other partner's use of social media. Who are they talking to online? Who are they texting? Very quickly partner monitoring, suspicion, jealousy, and mistrust manifest themselves around the spouse's contacts and relations with other people online, which can speed up marital breakdown and divorce. A decade ago, some researchers even predicted that social networking sites might contribute to an increase in divorce and infidelity rates due to the extent and ease of accessibility to connect with past partners [27]. Websites like Ashley Madison even offered opportunities for new partners even if you were already married, which probably didn't help people already worried about their relationship with their spouse. These days, there is even advice out there on the internet to help you detect if your partner is having a cyber-affair [28].

One domain where social media made a big impact was on how relationships start and as we just saw, how they can end. Some people were quick to spot a business opportunity tapping into people's desire to find love, lust or plain old human companionship. If you could connect with old girlfriends, then you could connect with new ones too. The world's first dating website began in 1995 with Match.com but the use of computers for finding love actually began almost 30 years earlier. In 1965, some Havard students, who were perhaps themselves unhappy with their own love lives, set up a business called Compatibility Research Inc. The enterprising students developed a paper-based

75-question survey plus additional questions on themselves (including their SAT scores) for love-hungry applicants to fill out. Individuals would then return their questionnaires by mail (along with a $3 fee) and the responses were fed into an IBM 1401 computer which was the computer equivalent of the Model T Ford as it was the world's first relatively affordable and widely popular mainframe computer [29]. The IBM would do its thing and individuals would receive a list of computer-generated matches in return. By 1966, the company claimed to have 90,000 individuals using the service which means that the students had made the equivalent of 2.5 million dollars today.

Today, love is even more lucrative. The number of people now using social media platforms to find a partner has ballooned since 1995. In 2023, the number of people using dating platforms like Tinder, Bumble and Hinge surpassed 321 million people and that number has been estimated to rise to 452 million by 2028 [30]. The revenue today is astonishing with almost 6 billion dollars [31]. The effect of these platforms has been dramatic and varied. One of the earliest observations that occurred shortly after the launch of Match.com in 1995 was a rise in interracial marriages and that trend has continued to rise alongside the development of different dating platforms. A decade ago, individuals who met via online dating reported more satisfying and stable marriages. Today, that trend seems to have reversed: people who met online and subsequently married now report their relationships are less satisfying and stable than those who met their spouse the old-fashioned way [32]. The explanations for the change may be due to the algorithms used to match people, or it could just simply be that the expectations that develop from the apparent size of the dating pool available is one of the factors that leads to differences in relationship outcomes. For example, when the options on dating apps seem abundant and easily accessible, people may be less willing to remain in a relationship when times get tough [33]. Of course, and just like other aspects of social media and networking platforms, the results for many have not been coming up cupid. Almost half of the users of dating platforms say they felt frustrated from online dating, with one-third reporting dating platforms made them feel pessimistic, and a quarter saying that using the platform made them feel insecure [34]. It can get a little worse too. Almost half of women (but also a quarter of men) on dating platforms say they were contacted on a dating site or app in a way that made them uncomfortable, with both men and women receiving unwanted inappropriate sexual messages or pictures. Finally, almost equal numbers of men and women claimed to have feared for their safety at some point, even going so far as to block someone completely or report them to the dating site [35].

It's also worthwhile being wary of those you swipe right on your profile. If you've never been a victim of a romance scam, consider yourself lucky. The basic idea is simple but often very effective. The scammer creates a fake profile with dazzlingly attractive pictures and personal attributes on the dating app and makes victims think they want a relationship, but the real objective is to get money, usually under the pretense that they can't continue the relationship without financial assistance, or they need it to come and physically meet you. Once a victim does fall for deceptively good looks on the profile, scammers quickly move to more direct messaging using WhatsApp, Telegram or SMS messages to carefully grow a personal relationship with their victim using well-rehearsed tactics to gain their trust and affection. These scams cost users a total of $75 million in 2016, a figure that grew to a whopping $304 million in 2020 [36]. If you think you're immune to falling for a romance scam, keep the following in mind: Scammers are now using AI-generated images to create unique dating profiles. It's a clever trick because anyone with a bit of tech savviness can conduct a quick search online to find duplicates of an image. But using an AI-generated picture helps scammers avoid skepticism from partner seekers who exercise some degree of necessary caution [37]. Sadly, but predictably, people searching for partners are now effectively scamming others by putting their own pictures through AI image-enhancing apps. Sadly, but predictably, these AI-enhanced pictures of yourself outperform the real ones in terms of attracting attention from potential suitors [38]. The temptation to use such apps in the face of the competition is high but it creates illusory expectations for everyone involved. It's likely that first dates for those initially meeting online are going to be increasingly disappointing in the future.

SEXUALITY AND THE INTERNET

The internet has also altered behavior and expectations surrounding one of the more intimate aspects of human behavior, sexuality. Of the 50 most visited websites on the internet, six of them are 'adult' or porn sites [39] and while it's estimated that only 4% of all pages on the internet are pornography related, the relative impact is over-sized. Three of those 50 sites account for almost 6 billion site visits every single month which is approximately 135,000 visits every single minute. By the way, the average length of a visit is 18 minutes. Almost 13% of all internet searches are for pornography content and over one-third of all internet downloads have some connection to adult material [40]. Some interesting patterns emerge with respect to internet-based pornography. Users' favorite time to watch sexually explicit content is in the small hours of the morning. Friday is the least favorite 'porn watching' day, while Sunday has the highest number of visitors [41]. A survey recent uncovered some startling insights about when people first encounter adult content. Three-quarters of individuals report having seen online porn by the age of 15, with just over one in ten admitting to having encountered it before reaching the age of 9 [42]. The internet has resulted in unprecedented levels of access to pornographic material and that's not without consequence. At a basic level, consistent exposure to pornography affects individuals' expectations regarding sexual activity, which in turn negatively affects their ability to form and maintain romantic, or sexual, relationships. It also seems to be the case that men who use pornography more frequently report less desire for their partner, and for sex in general and rates of sexual dysfunction have been increasing in younger age males [43]. It's not just an exclusively male activity either. The number of women visiting adult sites has been increasing and up to a third of women say they regularly use the internet to use the same free access sites that men do [44].

CONTENT MODERATORS: WHO PROTECTS THE PROTECTORS?

As a final issue, you might wonder who patrols the internet to try and keep it safe for users. Twenty years a wonderfully inventive and funny TV series aired on British TV hosted by an English national treasure, Tony Robinson. In each episode, Tony selected and educated viewers about truly awful historical occupations such as the leech collector, plague burier, ratcatcher and even the sin eater. He often tried his hand at these jobs with predictably hilarious results. Contemporary gun enthusiasts may not often wonder where early forms of gunpowder once came from but in the past the early gunpowder business in Britain relied on what were known as the Saltpetre-Men. These men had the task of digging up urine-soaked oil, latrines and chicken coops to get the vital ingredient for gunpowder, potassium nitrate. Apparently, well-rotted soil was the best type, and their days would be spent digging this stuff up. However, what made the job particularly unpleasant that to determine where to dig, and how deep, a Saltpetre Man had to taste the soil and manure to assess quality [45]. Unsurprisingly, a common complaint of people in the job was that they could not get rid of the smell. If you think that the era of awful occupations is well behind us, please spare a thought for one of the occupations spawned by the rise of the internet and social media, that of the 'content moderator'. The stuff they must deal with is way beyond the work of the Saltpetre-Men.

The walls of social media platforms must be kept nice and clean for the millions of people who use them every day. We expect them to be safe in the way we expect the television station not to show a porn movie after the lunchtime news on a Sunday when the children are home. Yet the problem is the primary characteristic associated with social media sites; that users can post content and comments. This includes posting the dark stuff. The beheadings, murder, assault, suicide, animal abuse, revenge porn, child abuse, racism, and extreme propaganda. To stop this material from appearing, social media platforms rely on a process called 'content moderation' which is basically a process for detecting content that most would consider obscene, illegal, harmful or insulting before they land on your preferred platform or at least very soon after. How do they do this you may well ask? The major platforms use a combination of algorithmic tools, user reporting and what is called human review by a content moderator. The job title content moderator sounds opaque and innocuous and stating your occupation at a party is likely to result in people asking you what does that involve? Chances are a content moderator won't be able to tell them for a variety of reasons including very tight confidentiality agreements but more probably because you just don't want to think about the job. These workers are the ones who decide whether or not content complies with the policies and guidelines of social media platforms. But they are also the ones who have to deal with the most brutal content: viewing it, evaluating it, censoring it and, if necessary, sending it to the police.

All social media platforms have what are called community standards, which as part of the drive toward greater transparency, sets out both content and decision-making standards for what appears on social media. Facebook, for example, has a specific set of rules for what is and what is not allowed on the platform. The company states the following: 'We remove content, even if it has some degree of newsworthiness, when leaving it up presents a risk of harm, such as physical, emotional and financial harm, or a direct threat to public safety' [46]. The platform also sets out prohibited content which is against its values across violence and criminal behavior (i.e., coordinating and promoting crime), safety (e.g., child and adult sexual exploitation), objectionable content (e.g., hate speech), integrity and authenticity (e.g., misinformation). However, irrespective of the stated standards, someone has to make sure that this material does not make it online and that's not easy given the numbers involved. Facebook has just over 2 billion daily users which means an awful lot of potential posts along with the opportunity for abuse to ensure content does not abuse the standards [47]. The funny thing about Facebook's commitment to values and standards is that it primarily relies on other companies to do the job of content moderation for them. Facebook (but also others) has a history of outsourcing the job of human content moderating to companies both in the USA and also in Africa [48]. The job of the moderator is simple: watch and read what is being posted. These individuals shape what is seen online, but also protect the reputation of the social media giants. Content moderators are often required to view depictions of child sexual abuse and violence and other content that displays cruelty, humiliation, racism and discrimination [49]. They can be exposed to this kind of content for hours per day. TikTok's content moderators are exposed to a regular diet of child pornography, rapes, beheadings and animal mutilation, according to a lawsuit filed against the video-sharing platform and its parent company [50]. One class-action suit has even cited footage of cannibalism. That's real Heart of Darkness stuff. You would think that they might be, given the job that they do, they would be among the group of employees best looked after. Except, they don't seem to be. Despite their importance to the business model of many online platforms, moderators are often undervalued, and expected to get on with it and are not afforded the same benefits and provisions as offered to other professionals. Several class action lawsuits have been launched in the USA, Spain and also in Africa and some have been successful. However, one problem is that syndromes such as vicarious trauma and burnout will typically occur cumulatively and gradually, which can make it difficult for individuals to recognize any changes within themselves or others. By the time you sit up and take notice, it might already be too late. Content moderators report a lot of negative psychological changes as a function of what they have to watch in order to keep others safe: Increased cynicism about people and the world, desensitization, anger, sleep disturbance, fatigue, hypervigilance regarding children and negative effects on relationships, and chronic anxiety [51].

BUT WHAT CAN WE DO?

It's hard to conclusively demonstrate the more negative aspects of social media usage among adults. People's lives are complicated, and disentangling the effects of the internet and social media on adult mental health is extremely difficult. Much of the research is correlational and proving causation is therefore always going to be tricky. However, thanks to Elon Musk's company, Starlink, we have had the chance to see what effect the internet has on a group of people who were previously untouched by it. Starlink's 2022 entry into Brazil made the high-speed digital connection possible for a 2,000-person tribe called the Marubo in a remote part of the Amazon. Everyone was excited at first about having access to the internet, but things started to unravel fairly quickly. While access to the internet has saved the lives of some members of the tribe because of better contact with medical services (snake bites are a big concern in the Amazon), the time devoted to hunting and fishing for survival diminished, younger men in the tribe started sharing pornography and displaying some unorthodox, sexualized behaviors and communication between tribe members suffered because many just spent hours on social media. They certainly seem to have caught up with life for the rest of us who have been exposed to the internet for the past

20 years. In response, the elders of the tribe decided to limit access to the internet to two hours each morning, five hours each evening, and all day Sunday [52]. They know it's useful, but they have seen first-hand and very quickly that it's really a modern-day Pandora's box. Perhaps there is a lesson in there for all of us.

REFERENCES

[1] "Jonestown," Encyclopedia Britannica. [Online]. Available: https://www.britannica.com/event/Jonestown [Accessed: Jun. 24, 2024].

[2] B. Bennett, "The strange life of reverend Sun Myung Moon," Foreign Policy, Sep. 4, 2012. [Online]. Available: https://foreignpolicy.com/2012/09/04/the-strange-life-of-reverend-sun-myung-moon/ [Accessed: Jun. 24, 2024].

[3] D. Chidester, *Salvation and Suicide: Jim Jones, the People's Temple and Jonestown* [Religion in North America], 2nd ed. Bloomington, IN: Indiana University Press, 2004.

[4] B. Beckett, "Missouri cult Rashad Jamal followers missing," The Independent, Jun. 24, 2024. [Online]. Available: https://www.independent.co.uk/news/world/americas/crime/missouri-cult-rashad-jamal-followers-missing-b2475283.html [Accessed: Jun. 24, 2024].

[5] A. Merlan. "An online prophet with a huge following has been convicted of child abuse," VICE, Aug. 21, 2023. [Online]. Available: https://www.vice.com/en/article/m7b99y/an-online-prophet-with-a-huge-following-has-been-convicted-of-child-abuse [Accessed: Jun. 24, 2024].

[6] "Most common reasons for internet use worldwide in 2022," Statista, Apr. 1, 2024. [Online]. Available: https://www.statista.com/statistics/1387375/internet-using-global-reasons/ [Accessed: Jun. 24, 2024].

[7] "Twin Flames Universe", Jun. 24, 2024. [Online]. Available: https://twinflamesuniverse.com/ [Accessed: Jun. 24, 2024].

[8] A. Romano, "Escaping twin flames: Inside the social media cult," Vox, Nov. 14, 2023. [Online]. Available: https://www.vox.com/culture/23959800/escaping-twin-flames-netflix-jeff-ayan-twin-flames-universe-cult-soulmates [Accessed: Jun. 24, 2024].

[9] M. Rousselet, O. Duretete, J. B. Hardouin, and M. Grall-Bronnec, "Cult membership: What factors contribute to joining or leaving?" *Psychiatry Research*, vol. 257, pp. 27–33, 2017. https://doi.org/10.1016/j.psychres.2017.07.018 [Accessed: Jun. 24, 2024].

[10] A. Lenhart. "Social Media and Friendships," Pew Research Center, Aug. 6, 2015. [Online]. Available: https://www.pewresearch.org/internet/2015/08/06/chapter-4-social-media-and-friendships/ [Accessed: Jun. 24, 2024].

[11] "Most common reasons for internet use worldwide in 2022," Statista, Apr. 24, 2024. [Online]. Available: https://www.statista.com/statistics/1387375/internet-using-global-reasons/ [Accessed: Jun. 24, 2024].

[12] P. Suciu, "Social media isn't really all that social anymore. Can it be again?" Forbes, Nov. 16, 2023. [Online]. Available: https://www.forbes.com/sites/petersuciu/2023/11/16/social-media-isnt-really-all-that-social-anymore-can-it-be-again/?sh=57d266c82bbe [Accessed: Jun. 24, 2024].

[13] E. C. Tandoc Jr. and Z. H. Goh, "Is Facebooking really depressing? Revisiting the relationships among social media use, envy, and depression," *Information, Communication & Society*, vol. 26, no. 3, pp. 551–567, 2023. https://doi.org/10.1080/1369118X.2021.1954975 [Accessed: Jun. 24, 2024].

[14] B. Goldsmith. "Facebook study says envy is rampant on the social network". HuffPost. Jan. 22, 2013. [Online]. Available: https://www.huffpost.com/entry/facebook-study-envy_n_2526549. [Accessed: Jun. 24, 2024].

[15] P. Reaney. "Rise in divorce evidence from social websites?" Reuters, Feb. 11, 2010. [Online]. Available: https://www.reuters.com/article/us-divorce-internet-idUSTRE6194ZE20100210/ [Accessed: Jun. 24, 2024].

[16] F. Carraturo, et al. "Envy, social comparison, and depression on social networking sites: a systematic review." *European Journal of Investigation in Health, Psychology and Education* vol. 13,2 364–376. Feb. 1, 2023, https://doi.org/10.3390/ejihpe13020027

[17] S. Kemp. "Digital 2024: The Time We Spend on Social Media," Datareportal, Jan. 31, 2024. [Online]. Available: https://datareportal.com/reports/digital-2024-deep-dive-the-time-we-spend-on-social-media#:~:text=Well%2C%20at%20an%20average%20of,years%20of%20collective%20human%20time [Accessed: Jun. 24, 2024].

[18] "Mental Health Challenges for Young Adults Illuminated by New Report," Harvard Graduate School of Education, Oct. 24, 2023. [Online]. Available: https://www.gse.harvard.edu/ideas/news/23/10/mental-health-challenges-young-adults-illuminated-new-report#:~:text=A%20perception%20that%20the%20world,others%20and%2034%25%20reported%20loneliness [Accessed: Jun. 24, 2024].

[19] A. M. Khalaf, A. A. Alubied, A. M. Khalaf, and A. A. Rifaey, "The impact of social media on the mental health of adolescents and young adults: A systematic review," *Cureus*, vol. 15, no. 8, e42990, 2023. https://doi.org/10.7759/cureus.42990 [Accessed: Jun. 24, 2024].

[20] "Adults' Media Use and Attitudes Report 2024," Ofcom, Apr. 19, 2024. [Online]. Available: https://www.ofcom.org.uk/__data/assets/pdf_file/0020/283025/adults-media-use-and-attitudes-report-2024.pdf [Accessed: Jun. 24, 2024].

[21] E. Calder. *Frequent internet use improves mental health of older adults*, University College London, Jul. 2020. [Online]. Available: https://www.ucl.ac.uk/news/2020/jul/frequent-internet-use-improves-mental-health-older-adults [Accessed: Jun. 24, 2024].

[22] M. J. Callan, H. Kim, and W. J. Matthews, "Age differences in social comparison tendency and personal relative deprivation," *Personality and Individual Differences*, vol. 87, pp. 196–199, 2015. https://doi.org/10.1016/j.paid.2015.08.003 [Accessed: Jun. 24, 2024].

[23] B. T. McDaniel, M. Drouin, and J. D. Cravens, "Do you have anything to hide? Infidelity-related behaviors on social media sites and marital satisfaction," *Computers in Human Behavior*, vol. 66, pp. 88–95, 2017. https://doi.org/10.1016/j.chb.2016.09.031 [Accessed: Jun. 24, 2024].

[24] "Facebook fueling divorce, research claims," The Telegraph, Dec. 21, 2009. [Online]. Available: https://www.telegraph.co.uk/technology/facebook/6857918/Facebook-fuelling-divorce-research-claims.html [Accessed: Jun. 24, 2024].

[25] J. Sisemore. "How does social media affect marriage?" Sisemore Law Firm, May 18, 2022. [Online]. Available: https://www.thetxattorneys.com/blog/how-does-social-media-affect-marriage [Accessed: Jun. 24, 2024].

[26] "How does social media affect relationships," Medical News Today, Feb. 15, 2023. [Online]. Available: https://www.medicalnewstoday.com/articles/social-media-and-relationships [Accessed: Jun. 24, 2024].

[27] R. McDermott, J. Fowler, and N. Christakis, "Breaking up is hard to do, unless everyone else is doing it too: Social network effects on divorce in a longitudinal sample," *Social Forces*, vol. 92, no. 2, pp. 491–519, 2013. https://doi.org/10.1093/sf/sot096 [Accessed: Jun. 24, 2024].

[28] S. Stritof. "Is your spouse having a cyber affair?" Verywell Mind, Dec. 28, 2022. [Online]. Available: https://www.verywellmind.com/is-spouse-having-a-cyber-affair-2300653 [Accessed: Jun. 24, 2024].

[29] "IBM 1401 Data Processing System," IBM Archives, Jun. 24, 2024. [Online]. Available: https://ww.ibm.com/history/1401 [Accessed: Jun. 24, 2024].

[30] S. Jo Dixon. "Online dating - Statistics & facts," Statista, Mar. 27, 2024. [Online]. Available: https://www.statista.com/topics/7443/online-dating/#topicOverview [Accessed: Jun. 24, 2024].

[31] A. Hadji-Vasilev. "Online dating statistics 2024," Cloudwards, May 19, 2024. [Online]. Available: https://www.cloudwards.net/online-dating-statistics/#:~:text=It%20is%20estimated%20that%20there,Does%20Online%20Dating%20Work%3F [Accessed: Jun. 24, 2024].

[32] S. Buck, "Unpacking the online dating effect," Psychology Today, Oct. 12, 2023. [Online]. Available: https://www.psychologytoday.com/us/blog/dating-in-the-digital-age/202310/unpacking-the-online-dating-effect#:~:text=Ten%20years%20ago%2C%20the%20direction,spouse%20the%20old%2Dfashioned%20way [Accessed: Jun. 24, 2024].

[33] L. L. Sharabi and E. Dorrance-Hall, "The online dating effect: Where a couple meets predicts the quality of their marriage," *Computers in Human Behavior*, vol. 150, pp. 1–8, 2024. https://doi.org/10.1016/j.chb.2023.107973 [Accessed: Jun. 24, 2024].

[34] E. Vogels. "10 Facts About Americans and Online Dating," Pew Research Center, Feb. 6, 2020. [Online]. Available: https://www.pewresearch.org/short-reads/2020/02/06/10-facts-about-americans-and-online-dating/ [Accessed: Jun. 24, 2024].

[35] J. Booth and O. Verhulst. "Dating statistics and facts in 2024," Forbes Health, Feb. 19, 2024. [Online]. Available: https://www.forbes.com/health/dating/dating-statistics/ [Accessed: Jun. 24, 2024].

[36] "Romance scams take record dollars in 2020," Federal Trade Commission, Feb. 2021. https://www.ftc.gov/news-events/data-visualizations/data-spotlight/2021/02/romance-scams-take-record-dollars-2020 [Accessed: Jun. 24, 2024].

[37] S. Pashankar. "Scammers Litter Dating Apps with AI-Generated Profile Pics," Bloomberg, Feb. 14, 2024. [Online]. Available: https://www.bloomberg.com/news/newsletters/2024-02-14/scammers-litter-dating-apps-with-ai-generated-profile-pics [Accessed: Jun. 24, 2024].

[38] S. Mueller. "I Put AI Photos on My Hinge Dating Profile, They Were the Most Liked by Far," Mashable, Oct. 26, 2023. [Online]. Available: https://me.mashable.com/sex-dating-relationships/34092/i-put-ai-photos-on-my-hinge-dating-profile-they-were-the-most-liked-by-far [Accessed: Jun. 24, 2024].

[39] "List of most-visited websites," Wikipedia, Jun. 24, 2024. [Online]. Available: https://en.wikipedia.org/wiki/List_of_most-visited_websites [Accessed: Jun. 24, 2024].

[40] K. Buchholz. "Share of the Internet that is porn," Statista, Feb. 11, 2019. [Online]. Available: https://www.statista.com/chart/16959/share-of-the-internet-that-is-porn/ [Accessed: Jun. 24, 2024].

[41] J. Wise. "How many porn sites are there?" EarthWeb, Mar. 21, 2023. [Online]. Available: https://earthweb.com/how-many-porn-sites-are-there/ [Accessed: Jun. 24, 2024].

[42] Pornhub Statistics, Scott Max, Jun. 24, 2024. [Online]. Available: https://scottmax.com/pornhub-statistics/ [Accessed: Jun. 24, 2024].

[43] A. D. Dwulit and P. Rzymski, "The potential associations of pornography use with sexual dysfunctions: An integrative literature review of observational studies," *Journal of Clinical Medicine*, vol. 8, no. 7, p. 914, 2019. https://doi.org/10.3390/jcm8070914 [Accessed: Jun. 24, 2024].

[44] F. Vera-Gray. "Yes, women watch porn: Here's what they're clicking on and why," The Times, Jan. 21, 2024. [Online]. Available: https://www.thetimes.com/uk/society/article/yes-women-watch-porn-heres-what-theyre-clicking-on-and-why-n059q59v5 [Accessed: Jun. 24, 2024].

[45] M. Pitt. "The 300 Industrial Secret That Changed the World," The Chemical Engineer, Sep. 29, 2022. [Online]. Available: https://www.thechemicalengineer.com/features/the-300-industrial-secret-that-changed-the-world/ [Accessed: Jun. 24, 2024].

[46] "Approach to Newsworthy Content," Meta Transparency Center, Aug. 29, 2023. [Online]. Available: https://transparency.meta.com/features/approach-to-newsworthy-content/ [Accessed: Jun. 24, 2024].

[47] B. Dean, "Facebook Users Statistics 2024," Backlinko, Feb. 23, 2024. [Online]. Available: https://backlinko.com/facebook-users [Accessed: Jun. 24, 2024].

[48] E. Hannon, "Facebook Content Moderators Call the Work They Do Torture, Their Lawsuit May Ripple Worldwide," Euronews Next, Jun. 26, 2023. [Online]. Available: https://www.euronews.com/next/2023/06/29/facebook-content-moderators-call-the-work-they-do-torture-their-lawsuit-may-ripple-worldwi [Accessed: Jun. 24, 2024].

[49] R. Spence, A. Bifulco, P. Bradbury, E. Martellozzo, and J. DeMarco, "The psychological impacts of content moderation on content moderators: A qualitative study," *Cyberpsychology: Journal of Psychosocial Research on Cyberspace*, vol. 17, no. 4, Art. 8, 2023. https://doi.org/10.5817/CP2023-4-8 [Accessed: Jun. 24, 2024].

[50] R. Burnson. "TikTok Sued by Content Moderator Traumatized by Graphic Videos," Bloomberg, Dec. 24, 2021. [Online]. Available: https://www.bloomberg.com/news/articles/2021-12-24/tiktok-sued-by-content-moderator-traumatized-by-graphic-videos [Accessed: Jun. 24, 2024].

[51] V. Sood. "Ex-cognizant staff sue over mental health harm from facebook work," Livemint, May 27, 2024. [Online]. Available: https://www.livemint.com/companies/news/excognizant-staff-sue-over-mental-health-harm-from-facebook-work-11716700739647.html [Accessed: Jun. 24, 2024].

[52] A. Court. "Remote Amazon Tribe Connects to Elon Musk's Starlink Internet Service, Become Hooked on Porn, Social Media," New York Post, Jun. 4, 2024. [Online]. Available: [Online]. Available: https://nypost.com/2024/06/04/lifestyle/remote-amazon-tribe-connects-to-elon-musks-starlink-internet-service-become-hooked-on-porn-social-media/ [Accessed: Jun. 24, 2024].

8 Pixelated Perils – The Double-Edged Sword of Online Gaming

In a world before smartphones, computers, and video games, long ago, humans got bored. Regardless of location or timeline, people always craved entertainment and one source of this was playing games. They used whatever was around them to get creative and come up with several activities. One of the earliest examples of a board game was what the ancient Egyptians called 'Senet,' which was their version of what we call 'Checkers' today [1, 2]. The game was for two players, which had a board of 30 squares arranged into parallel rows of 10 squares each. The supposed goal of the game was that two players would square off against each other, and they would try to move all their pieces to their opponent's side while preventing the other player from doing the same. Not to be outdone, the Romans and the Ancient Greeks, also had a variety of games, some of which have influenced some modern ones that we play today. One excellent example is a Roman game called Terni Lapilli, which is essentially the ancestor of the game we know as Tic-Tac-Toe [3]. Stepping away from boards and onto courts and fields, the Greeks had a ball game named Episkyros, where two teams had to throw a ball over each other to see who could cross the white line drawn behind their opponents first [4]. If you think about this for a moment, you'll quickly realize that this is essentially an ancient version of modern-day American football or rugby.

DOI: 10.1201/9781032679389-9

In civilizations further east, games that are still played today were also created. For example, a game called Go was played approximately 4000 years ago in China, Japan, and Korea [5]. In Ancient China, there was a game called Weiqi, which was so revered that it was considered to be one of the four cultivated scholarly arts, alongside calligraphy, painting, and playing a traditional musical instrument called the Guqin. Playing cards, dominoes, and Xiangqi (Chinese chess) were also all created in the region and are obviously games that are still played today.

PRESS START TO BEGIN

The point of all this is that games have existed for a very, very long time and exist because they satisfy some basic needs we have for fun and interaction with others. It's also the case that many of the games we play today, be they traditional tabletop games (board games, card games, etc.) and even contemporary video games, have been inspired by these ancient games. So how did we go from moving little pieces on boards to mashing buttons and staring at computer screens? The idea of digital gaming started in 1952, when OXO (basically Tic-Tac-Toe) was created as a part of a doctoral dissertation at the University of Cambridge [6]. But this was a privately created video game. Then a guy called Ralph Baer came along and he wanted to explore the possibility of playing digital games at home. Later dubbed the Father of Video Games, he created the first home console, the Brown Box, which led to the development of the consoles we have today [6, 7]. Eventually, in 1975, Pong, the first commercial video game, was released for the Atari console, and the digital gaming world picked up from there.

What followed was the golden age of arcade video games. You see, we didn't jump from simple little Pong straight into where we are today with spending hours gaming in the confines of our bedrooms. Actually, before proper PC and console gaming, came arcade gaming. These were gigantic machines enticing you with 8-bit tunes and big shiny colorful buttons. At the cost of 25 cents, you were transported into a very limited world where you could eat pellets while escaping ghosts or saving the princess from a giant gorilla. Think of other games like Frogger, Space Invaders, and Asteroids. An arcade was the hot spot to hang out when the 1980s rolled around. I wasn't alive yet, but from what's been depicted in movies and shows it looked like a great place to be, people gathering around giant consoles with pockets full of coins and cheering fellow gamers.

The time period between 1978 and 1983 was what we now call the golden age because that's when gaming popularity began to spike. It started with the release of Space Invaders in the USA in 1978, which grew in popularity over the next year, which coincided with the development of vector display technology used in arcade consoles [8, 9]. By 1980, the US arcade industry had generated approximately $2.8 billion... all from quarters [10, 11]. Two years later, in 1982, the revenue from the arcade industry, again from quarters, had grown to almost $8 billion [11, 12]. Despite that success, by 1983, the arcade video game market had begun to stagnate and one reason for this was because home video game consoles started gaining popularity. Video gaming didn't go extinct, it began to evolve.

Arcades, during this period, had started holding competitive tournaments to further improve business, and major gaming companies like Sega got in on the action by sponsoring some of these tournaments themselves [13]. Competitive arcade gaming was even televised and presented on shows like Starcade in the United States or First Class in the United Kingdom [13]. Actually, even now, we still have retro (old-style arcade) game competitions with games like Tetris, Mortal Kombat, or Final Fight. The tournaments that took place back then were considered esports. Esports, after all, does stand for 'electronic sports,' and involves professional or amateur players or teams competing against each other in video games [14].

Today, the esports industry is MASSIVE. We're talking lots of tournaments around the world (local and international, big and small), and it's a multi-billion-dollar industry, and lots of cash to win for participating teams. As of March 2024, the worldwide revenue of the esports industry was $4.3 billion (for reference, the industry revenue in 2023 was $3.8 billion, and before that, it was $3.2 billion in 2022) [14]. In 2024 alone (up to June 2024), some of the most popular games have already grown their prize pool to millions of dollars. So far (at the time of writing this, it's mid-season for most competitive games), the game Counterstrike has an $8.7 million prize pool, DotA 2 has almost $7 million, Rainbow Six Siege has $5.3 million, Fortnite has $4.6 million, and League of Legends has $4.2 million [15]. While these prize pools are currently in the millions, much of the income comes from merchandise, game developer fees, sponsorships, and streaming.

The cool thing about anything sports related, not just esports, is that you don't need to go to the venue to watch these tournaments. You can watch them from the comfort of your home. Well, almost-comfort. The number of times my friends and I would stay up late just to watch the world final of our favorite games is too damn high! But streaming is actually one of the ways esports teams get their viewership in the first place. More viewership means more success in the future (from sponsorships, investors, deciding if the whole tournament is worth hosting in the first place, deciding if the game is worth supporting (i.e., if people like it), etc.). In January 2024, StreamElements, one of the big companies behind streaming tools and services, reported that TwitchTV had a total of 1.902 billion viewership hours [16]. In case you don't want to do the math, that's approximately 7.925 million days. DAYS. All of which were watched in the span of one month.

Now that you know what esports are, let me backtrack a little to the idea of social gaming. All the games mentioned so far are in fact social games, because they involve multiplayer activities with objectives for players to complete, either together or against one another [17, 18]. Let me pick the one I play as an example: League of Legends. This game involves two teams, and you must work together with your teammates to defeat the enemy team by reaching their base (a little bit like Senet!). Of course, to do this, you need to strategize and communicate with your allies – in other words, to socialize with them. You see, the cool thing about video games is that in today's world, they're not all offline solo-play games. The internet has become so much more accessible, and more developers are integrating network-based components to video games. This enables players to socialize with others within the game and even play with them, regardless of their location around the world. Theoretically, this is great. People can socialize more from the comfort of their home, and they won't be limited to whoever is around them.

Online games (those played over the internet) have been steadily growing in popularity over the years, as you can see from Figure 8.1. In 2023, there were an estimated 1.13 billion online-specific gamers, and that number is projected to grow to 1.17 billion by the end of 2024 [19]. By the way, a 'gamer' is someone who regularly spends at least a couple of hours a day

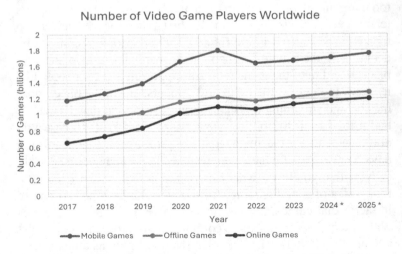

FIGURE 8.1   Ten-year time frame of the number (in billions) of gamers (online, offline, and mobile) worldwide [19].

playing video games. If you look at the figure again, you can see the numbers for mobile and downloaded (offline) games follow almost the same pattern as online games. More and more people just keep turning to video games, which begs the question: Why are people playing in the first place?

HIGHS AND LOWS OF GAMING

Well, there are several reasons. The first is simply leisure or entertainment purposes – it's fun. But there are other reasons why people would want to play video games. In a 2024 survey of gamers conducted in the US by the Entertainment Software Association, [20] over two-thirds of the respondents said it was to 'pass the time or relax,' and an equivalent number said they played games to have fun. Other reasons that popped up included 'immersion and escape,' connecting with friends, family, and community, competition, and inducing a feeling of accomplishment. Almost three-quarters of adults in the survey said they also believed that video games were good for teaching problem-solving skills and two-thirds reported they were good for improving teamwork and collaboration [20].

It's also worth pointing out some other benefits too. Did you know video games can be extremely helpful in healthcare? And I'm talking about both physical and mental health. For physical health, video games like DanceDanceRevolution, Wii Fit, and Ring Fit Adventure have been shown to help with weight loss and physical therapy [21–23]. When it comes to mental health, video games appear to have therapeutic effects for individuals with symptoms of depression or anxiety. For attention disorders, like ADHD, the Food and Drug Administration in the US have authorized a game called EndeavorX, which is tailored for children with ADHD [24], which can help improve motivation and concentration through incentivization and can be more cost-effective than medications [25–27]. Video games can also have educational benefits! It sounds weird, I know but research has shown that incorporating games into educational activities can help with motivation and learning for math, languages, sciences, and vocabulary [28, 29]. Again, incentivization is great; in order for the kid to progress through the game and get rewarded, they have to complete the tasks, and in order to do that, they have to complete the lessons being taught (presented to them in a more entertaining format, of course). There was even a whole Simpson's episode about it! You know Bart, troublesome kid, always causing problems at school, home, around the town. Well, the town had come into a lot of money and Marge managed to convince (with John Legend's help) the townsfolk to open a science- and technology-based school. Bart, upon realizing that he was being a tailored game-based education, was so excited that he was completing all his assignments on time and appropriately to unlock skins (different outfits) for his school's digital avatar. Again, incentivization is great.

You might think at this point that video games are all fun and games and there are no downsides but just like we have seen in other areas of life on the internet, it's not that simple. In fact, the Klaxons were sounding about the perils of video gaming a lot earlier than you might think. In 1982, Space Invaders was all the rage and in the same year some psychiatrists in the USA published a paper about the rather unusual behavior of three men aged between 25 and 37 years who played the game and, as it just so happened, were also about to get married. One man began taking his fiancée on 'dates' to play or watch him play the game and one of the men even delayed his honeymoon excursion for an hour to get in a few more games. I suspect neither of their wives were entirely pleased with them in those moments. The really interesting point, though, was that all three men had QUADRUPLED the frequency of their playtime (up to 15 times per week), and all reported fantasizing about the game before falling asleep at night. For the first time, obsession had appeared in the context of video gaming [30] and it hasn't gone away. It's easy to see why when you look at the sheer amounts of time that some people spend on video games. One survey conducted in the US toward the end of 2023 reported that 22% of individuals in the 18-to-29-year age bracket were spending 6–10 hours per week playing video games and 8% were devoting more than 20 hours to gaming in an average week [31]. When you realize that over 200 million people in the USA are considered regular gamers, those figures add up to a substantial amount of time [32].

However, the amount of time spent doing something you like does not automatically indicate the presence of a problem. After all, people devote lots of time to other activities like cooking, hobbies and going to the gym and that's not typically a problem. But for some people, video gaming can become problematic or disordered. It's hard not to argue there is a problem when personal hygiene becomes neglected, there is significant weight loss or gain, excessive playing results in sleep deprivation, they play at work, and avoid contact with friends to play, or end up lying about how much time they spend playing video games. Many argue that this profile of dysfunction puts it on a par with other addictive behaviors.

In fact, the World Health Organization, which publishes the ICD-11 (the International Classification of Diseases) has included Gaming Disorder as a psychiatric condition and defined it as a lack of control over gaming and an extreme preoccupation with video games to such an extent that it takes priority over the regular activities of life such as work, socialization [33]. ICD's American cousin, the DSM-5 published by the American Psychiatric Association hasn't gone quite this far but has stated that it's a condition that is in need of further study. One study conducted in 2022 looked at the gaming patterns of over 400,000 individuals from 155 studies across 33 different countries. The study found that only 10% of children and adolescents (aged 8–18) and young adults (aged 18–28) met the criteria for disorder gaming, which is a relatively small amount given the overall number of people playing [34]. But the interesting thing was they also looked at risk factors for problematic gaming. Risks included stress, low self-esteem, emotional distress, depression, anxiety, interpersonal conflicts, being victim of bullying, poor academic performance, hyperactivity or inattention, and the presence of family dysfunction. In other words, too much gaming is, in some ways, just a means of coping with negative stuff going on in their lives. Maybe not the best way but still a way. Excessive gaming can also have some negative physical health effects too. If you've ever seen stereotypical gamer nerds in TV shows or movies, you've likely also seen them either massaging their hands and wrists or wearing a wristband. Well, they do that because one of the most well-known physical effects of too much gaming is Carpel Tunnel syndrome, where the wrists inflame and feel cramped. Another common issue is something that has been dubbed 'Gamer's Thumb,' where the thumb becomes inflamed [35]. This is seen mostly in players who frequently use handheld controllers to play video games. These two hand-related conditions are due to the strain put on the muscles in these body parts while keeping them constantly in the same position and ready for action.

Another risk that online video gaming can expose you to is harassment and verbal abuse. Among the most popular video games that people play are competitive PvP (player versus player) ones, such as League of Legends, DotA 2, Overwatch, and Fortnite. Now, these games don't all belong to the same genre or type, but they do seem to attract an extremely competitive, sore winner/loser type pattern of behavior. Not everyone has to agree with the ways and whys of someone else playing in their game, especially in team-based multiplayer games. However, some particular individuals will use this as a reason to harass other players in-game (and sometimes out of it, by continuing the harassment via direct messages (DMs) to the person or, in more extreme but thankfully rare cases, stalking). While the quality of the insults used in some contexts is laughable (e.g., 'X YOUR MOM'), harassment can have some effects on the mental well-being of the person on the receiving end, especially if they are extremely disturbing or violent. Figure 8.2 presents two examples of 'funny' harassment (where the funny part here is that it's a stupid little figure) and extreme harassment (be warned: the language in the extreme example is very harsh).

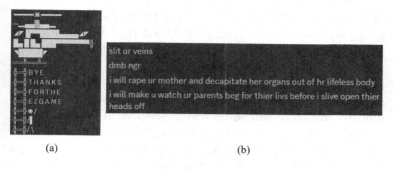

(a) (b)

FIGURE 8.2 Examples of (a) funny and (b) extreme harassment.

Every year, since 2019, the Anti-Defamation League (ADL) has conducted a study in the US to look at the extent of in-game harassment for both youth and adults. They not only look at general harassment, but they also look at identity-driven attacks, exposure to controversial topics, and the impacts on players. Their study found that 74% of minors (ages 10–17) and 76% of adults (aged 18 and older) have experienced some form of harassment in the latest survey from 2023, and its increasing year on year [36]. There are different types of harassment that one can experience in-game too, not just your usual 'you're shit, go kill yourself,' which is verbal abuse. You also have hate speech (abuse targeted toward specific identities), and threats against someone's life or describing potential violent acts against them [36].

If gamers experience anything that upsets them in-game, they are given the option to report it. This can be something as obvious as language or ambiguous as suspicious or annoying play behavior. When you make an account for any game, you are presented with a code of conduct or terms of service where the company will explicitly mention various negative behaviors and force you to accept their terms before you can even play. If you are caught behaving badly, you will likely be punished. The extent of the punishment depends on the extent of the 'crime.'

Punishments can range from simple warnings about the language you used to full-on account bans. Companies do try their best to prevent malicious behavior from occurring in the first place. But some perspective is needed. Between 3 and 3.2 billion people are considered gamers, with Asia taking up half of that number [37]. Now, there are thousands of games out there, some globally available and some region-specific. If we dig deeper and look at this from a slightly more technical perspective, try to think of all the servers that network traffic has to constantly go through. How does the abuse get picked up and stopped?

Game moderation is a difficult task in more ways than the obvious (making sure the rules are being followed). What complicates the task is that gaming jargon is just like any other language; you won't understand it if it's not familiar to you. Relying on human moderators to monitor game content shared online (chats, visual media, links) makes the whole process prone to human error. If the bad language used was in Portuguese, for example, but the mod handling the report ticket only speaks English and French, they will either pass on the ticket to someone else in hopes that they understand the language or disregard the report altogether (which happens often, due to the sheer volume of content that they have to sift through in the first place) [38, 39]. This is where we can incorporate AI to help.

AI moderation can assist humans by reducing the amount of potentially disturbing content they have to review, but even these automated methods have their limitations. For one thing, AI moderation is also at risk of human error; after all, it is a team of humans that developed, trained, and deployed the system. The other thing to worry about is the ever-evolving language. Unlike written languages, gaming language tends to evolve similarly to social media language. For example, my generation used acronyms like GTFO ('get the f* out'), and BM ('bad manners') Nowadays, however, kids use phrases like GOAT ('greatest of all time,' used for players that have done something worthy of compliment), yeet (to throw something), and bussin' (used to describe something that is really good). The implication is that a language that keeps on changing is a language that will be hard to track. So, even if we have datasets of game chats to train AI models, if the language evolves, then the model will need to continue being retrained if it doesn't manage to adjust and learn these phrases on its own. This depends, again, on the way it has been implemented, which is done by humans, who are forever prone to error. So, language that uses new terms that are unfamiliar to both the mod and the AI model will be overlooked and can avoid being filtered unless it's bad enough that plenty of reports have been made.

The players know this, too. And even if they don't understand how the reporting system works, they'll try to avoid the filters anyway. After all, where there's a will, there's a way. 'What can I say that won't get me banned,' is what they'll be asking themselves. Let me pick one out to use as an example, 'GC KYS GAD.' Just looks like a string of letters, right? At first sight, they don't necessarily make any sense, but they have grown to mean something in gaming language. These are actually abbreviations of phrases that would otherwise trigger the game's chat automod. GC = Get Cancer, KYS = Kill Yourself, and GAD = Get a Disease. Makes more sense, now, doesn't it? You'd hate to have someone tell you these things, right, especially if they hit close to home. Alas, unless they were added to the filter list or the AI was trained to know their meanings, these abbreviations just look like random strings because they are not 'common' abbreviations (for example, everyone knows what LOL or WTF means).

Unfortunately, there are ways to get around the rules by using abbreviations, symbols, and ambiguous language. It's important to know how to avoid or handle encounters with this kind of behavior, be it through your own filters (including mental ones) or using those available to you (most games have the option to mute chat altogether, so you don't have to communicate with anyone). Now, if you're a parent, this can be difficult to manage. Sure, you can do this without your kid knowing, but if they know enough about the game and technology, they'll find a way to re-enable the chat or work around filters. Where there's a will, there's a way. You can always be the super concerned parent and install parental moderator tools for the computer, but that can have a few consequences: (1) You'll possibly lose the respect your kid has for you, (2) a lot of these tools are really just a cover for malware, and (3) the game's built-in anti-cheat tool might see the parental software as a third party application and either ban your child's account or prevent the game from opening.

BOSS-LEVEL THREATS

Game chat moderation to stop abuse is a constant battle in online gaming, but there are other things to worry such as malware and plain old theft. Now, games are usually developed in secure environments and checked for security flaws before deployment. Developers will usually do a good job, but again, human error is a thing. Different components of a game with online features can be abused. One of the more common examples of this type of abuse is through RCEs (Remote Code Execution), where an attacker can remotely run malicious scripts (little snippets of code) to target an individual. How do RCEs get an opportunity for attack in the first place? Well, one way is through game add-ons. Some games, like World of Warcraft, allow users to install add-ons that can improve the quality of some aspects of the game, such as offering a cleaner, customizable user interface, or keeping track of and announcing upcoming enemy abilities. Players can also share scripts with one another via in-game links. One such script was used to exploit the game's auction house system to take 'gold' (in game currency) from other players. The script would overprice the item to be sold while making the auction house entry in the game appear legitimate to the victim [40, 41]. Once the item was sold, the victim would notice that they actually lost significantly more gold than what they thought was listed.

If it's not the game itself being used as a medium for attack, malicious users can take advantage of a player's desire to reach higher levels within the game or to enhance their overall game experience. In March 2024, a user named Cartis Tweaks conducted a malicious campaign on YouTube to attract users to join a Discord channel related to the game Roblox, which is commonly played by kids and influencers [42, 43]. They then took advantage of youthful innocence (or desperation) and encouraged viewers to download a PC optimizer that would help them play the game better. Unfortunately, this is the equivalent of 'downloading more RAM' to increase the memory of your PC because downloading more things will likely reduce the quality of a game rather than improve it. In this case, the thing that they downloaded was called a 'Tweaks' or 'Tweaker,' which is malware designed to steal information. Once downloaded, this malware would run in the background of the victim's PC and begin uploading sensitive information (like username and password combinations and the services they are used for) and stream it to the criminal's command and control center.

Enticing someone to download malware to 'improve' a game's quality is just one way to try to scam someone. In case you haven't encountered these things yourself, many online games, especially MMORPGs (mass multiplayer online role-playing games), have their own virtual economy. One example is World of Warcraft, which has features like trading resources, supply and demand, and strategic sales. Naturally, if you struggle to make (in-game) money, you might struggle with playing the actual game, especially toward the 'endgame,' where you are expected to be at the maximum level with the appropriate gear to play more advanced content. One thing that these games like this are known for is the presence of boosters and gold sellers, people that offer in-game services such as boosting you through chunks of content or supplying you with loads of in-game gold. All for a price, of course. Many boosters will use in-game currency, but some will go so far as to demand real money. Of course, these services have their risks. For one thing, they could be scams, just like Tweaker, except instead of infecting you with malware, they'd be stealing your (real) money [44]. Granted, boosters could attempt to scam you further by making it a requirement to input potentially malicious scripts. The other risk is that these services often go against a game's terms of service, and you could have your account banned for trying to participate, either as a client or a provider [45].

ROLLING THE DICE

There is another risk with games that may seem initially quite harmless but in fact is not. If you've played any modern game, be it on PC, console, or mobile, perhaps you've encountered what is called a 'lootbox system.' Sometimes, they're not directly called lootboxes and might be called summons, wishes, recruitments, reward boxes, card packs, treasure chests, whatever. Whatever the name, they all have the same characteristics. They are collected in the game and can be redeemed to receive a randomized selection of further virtual items, or loot, ranging from simple customization options for a player's avatar or character to game-changing equipment such as weapons and armor. They are attractive to players because they can enhance in-game quality. However, the key thing here is chance or luck. For example, you might have a 55% chance to receive 'common' loot, a 35% chance to receive 'rare,' an 8.5% chance of receiving a 'legendary,' and a 1% chance to receive an 'epic/mythic/godly' or whatever the game developers have decided to call the highest tier of loot. The remaining 0.5% chance is to gain something commonly known as a 'banner exclusive,' which is a special temporary character/item that is presented as part of an event with no other way to earn the prize. These are usually available for only a short amount of time and have a very low chance of reappearing again later (depending on the game hosts, of course).

Loot boxes belong to what is known as a 'gacha' system. 'Gacha' is short for gachapon, which was originally a physical capsule-earning 'game' popular in Japan. The idea is that you play a minigame or input some amount of money into the machine to get a tiny capsule. Within it, you get a prize of some kind just like how a Kinder egg operates. The quality, type, and amount of the prize depends on the game being played and the amount of money spent. It could be as simple as inputting 25 cents to get a capsule containing 1 of 6 different little keychains to paying $20 for a box containing a variety of special toys, clothes, snacks and/or collectibles. Game developers hit on this idea to add in-game incentives for players to continue playing the game while encouraging them to try to get features that enhance game quality. You would think this is a win-win situation; the companies maintain their game populations while the gamer is entertained and keen to have a better-quality game. Except the issue is that it's actually gambling and gambling can quickly turn into a problem.

You can buy these boxes using the in-game currency that you earn as you play. You can also spend money. Like, real-world money. And sadly, some people do. If you look at the percentages earlier the odds are clearly against you, right? How many boxes would you have to buy to guarantee that you get the exclusive reward? It also gets more complicated. In some games, it's not enough to simply claim the character/item once. You might want to 'roll the dice' multiple times because having multiple copies of the same character/item will allow you to level up and make the game experience better. Now you might be more interested in rolling more and more. After all, if you're going to try so hard to claim the prize, why leave it to simple ownership? Why not make it stronger? What if the game character/item is so weak on its own and NEEDS to be leveled up in order to be playable? Congratulations, now you have taken your first steps into the world of gambling. Gambling is simply the practice of risking money or other stakes in a game or bet [46] and to access lootboxes in games, you are risking either virtual or actual money. Therefore, lootbox systems are in effect gambling systems. Several studies have found links between lootbox systems and problem gambling in both adolescents and adults in several countries [47–50]. Some countries, like Belgium, the

Netherlands, the US, the UK, China, and Japan consider loot-boxes as a form of gambling with associated legal regulations.

I wish I could say that this issue is just limited to just loot-boxes. Many games have services known as 'microtransactions,' where you spend real money for little in-game rewards or boosts. These can be battle passes/milestone rewards or 'lucky discounts' for extra in-game currency or items that might boost your account faster than if you were to remain F2P (free-to-play). One example is when you start a new mobile phone game that has some kind of ranking system, and you get a pop-up once you finish the tutorial with a 'discount' on a 'welcome package.' Usually, this entails spending $0.99 for a mini bundle of extra in-game currency, including the real-world money-linked one (most of the time, games will have two currencies: the one that you freely earn as you play the game and the other that you have to buy). Now, $0.99 doesn't sound like much, right? But imagine you keep falling for this trap. Or you consistently buy $5 battle passes, which do already have free rewards, but you need to spend real money to get the better bonuses. That kind of spending stacks up.

Unfortunately, kids are also falling into these gambling and microtransaction holes to get the best loot possible. One case of this was an 11-year-old boy that spent £464 in one week on card packs for a mobile game [51]. The kid said that they did it because the game got better after they started buying the packs. A more extreme case is the 13-year-old who spent $64,000 of her parents' money over 4 months for in-app purchases [52, 53]. One might argue that she was young and got baited by the games and didn't understand what she was doing but she also deleted the transaction records to cover her tracks. Countries around the world are beginning to treat these systems as gambling and have created new laws and regulations, especially to protect players younger than 18 years of age. Japan and China were the first countries to attempt to crack down on gacha systems and micro-transactions. In 2012, Japan first attempted to crack down on gachas by banning 'complete' or 'compu' gachas, where you roll for pieces of the final prize instead of the whole prize itself [54, 55]. This tempted users to keep buying boxes in hopes of *finally* completing the final reward. Except, of course, there was always the chance that you never get that final piece and instead keep getting repeats of the pieces you already have, meaning that you essentially gain nothing but addiction and desperation while losing a lot of money.

Some countries have pushed game developers to disclose to gamers what exactly they were rolling for, the precise chances of success and other items that may be included in the lootbox [55, 56]. China first implemented this, which was done to give the player a chance to think about whether they want to try their luck to roll for the item or not, whether that's done using in-game or real money. While Asian countries are trying to regulate loot-boxes and microtransactions, European countries are taking harsher stances on the system. Belgium, for example, has created a law stating that lootboxes are illegal and that those involved could be fined and even imprisoned [57]. Punishments are also doubled if minors are involved in any way. In the UK, lootboxes and gacha systems are unavailable to minors unless they are accompanied by a parent or guardian to approve their purchases [58].

Video games still have their magic though, whether you play them online or offline. The thrill of exploring fantasy worlds or strategizing how to defeat the enemy team and defeat their nexus. It's great. It's relaxing. It might even sound a little absurd. But I guess it has something to do with being able to think of something else for a change, something that gives you more direct results. You enter a game and finish objectives after 30 minutes, just a couple of hours. Real-world tasks are long-term goals, and video games often provide short-term accessible achievements. In moderation with plenty of awareness, it's a great escape from the realities that we live in. Just watch out for those lootboxes.

REFERENCES

[1] J. Mark, "Games, Sports & Recreation in Ancient Egypt," World History Encyclopedia, Apr. 11, 2017, [Online]. Available: https://www.worldhistory.org/article/1036/games-sports--recreation-in-ancient-egypt/. [Accessed: Jun. 25, 2024].

[2] P. Piccione, "In Search of the Meaning of Senet," Archaeology, Jul. 1980, [Online]. Available: https://www.semanticscholar.org/paper/In-Search-of-the-Meaning-of-Senet-Piccione/bde0a8cf498b6e1d1a280074d2e5f300a8158998. [Accessed: Jun. 25, 2024].

[3] "Terni Lapilli," Tabletopia. [Online]. Available: https://tabletopia.com/games/terni-lapilli. [Accessed: Jun. 25, 2024].

[4] D. Elmer, "Epikoinos: The Ball Game Episkuros and Iliad," *Classical Philology*, vol. 103, no. 4, pp. 414–423, Oct. 2008, [Online]. Available: https://doi.org/10.1086/597184. [Accessed: Jun. 25, 2024].

[5] Britannica, "go - game," Britannica, Jun. 24, 2024. [Online]. Available: https://www.britannica.com/topic/go-game. [Accessed: Jun. 25, 2024].

[6] "Video Game History," History.com, Oct. 17, 2022, [Online]. Available: https://www.history.com/topics/inventions/history-of-video-games. [Accessed: Jun. 25, 2024].

[7] "The History of Video Games," Culture of Gaming, Dec. 22, 2022, [Online]. Available: https://cultureofgaming.com/the-history-of-video-games/. [Accessed: Jun. 25, 2024].

[8] J. Whittaker, *The Cyberspace Handbook*, Routledge, Dec. 4, 2003, [Online]. Available: https://doi.org/10.4324/9780203486023. [Accessed: Jun. 25, 2024].

[9] S. Kent, "The Ultimate History of Video Games," United States, Oct. 2001. [Accessed: Jun. 25, 2024].

[10] "Electronic Education," Electronic Communications, Incorporated, 1983. [Accessed: Jun. 25, 2024].

[11] E. Johnson, "Video Game Sales: 1972–1999," Gaming Alexandria, Jun. 7, 2021, [Online]. Available: https://www.gamingalexandria.com/wp/2021/06/video-game-sales-1972-1999/#arcadescoin-op. [Accessed: Jun. 25, 2024].

[12] E. Rogers and J. Larsen, "Silicon Valley Fever," Basic Books, 1984. [Accessed: Jun. 25, 2024].

[13] Rohan, "When did Esports start? When did esports become popular? The History of esports," Esports GG, Dec. 30, 2022, [Online]. Available: https://esports.gg/guides/esports/the-history-of-esports/. [Accessed: Jun. 25, 2024].

[14] "Esports - Worldwide," Statista, Mar. 2024, [Online]. Available: https://www.statista.com/outlook/amo/esports/worldwide. [Accessed: Jun. 25, 2024].

[15] "Top esports games in 2024 by prize money," Esports Charts. [Online]. Available: https://escharts.com/top-games. [Accessed: Jun. 25, 2024].

[16] "State of the Stream for Jan 2024: Twitch has 5 months of daily growth, Palworld makes a stellar debut," Medium, Feb. 13, 2024, [Online]. Available: https://blog.streamelements.com/state-of-the-stream-for-jan-2024-twitch-has-5-months-of-daily-growth-palworld-has-stellar-debut-61049f6f002a. [Accessed: Jun. 25, 2024].

[17] J. Stolz, "The theory of social games: outline of a general theory for the social sciences," Humanities and Social Sciences Communications, Jun. 30, 2023, [Online]. Available: https://www.nature.com/articles/s41599-023-01862-0. [Accessed: Jun. 25, 2024].

[18] "What exactly are social games and what makes them different?," Medium, Sep. 17, 2018. [Online]. Available: https://medium.com/buff-game/what-exactly-are-social-games-and-what-makes-them-different-e24623382558. [Accessed: Jun. 25, 2024].

[19] "Number of digital video game users worldwide from 2017 to 2027, by segment (in billions)," Statista, Jan. 2024, [Online]. Available: https://www.statista.com/forecasts/456610/video-games-users-in-the-world-forecast. [Accessed: Jun. 25, 2024].

[20] "2024 Essential Facts About the U.S. Video Game Industry," Entertainment Software Association, 2024, [Online]. Available: https://www.theesa.com/resources/essential-facts-about-the-us-video-game-industry/2024-data/. [Accessed: Jun. 25, 2024].

[21] C. Comeras-Chueca, J. Marin-Puyalto, A. Matute-Llorente, G. Vicente-Rodriguez, J. A. Casajus, and A. Gonzalez-Aguero, "The effects of active video games on health-related physical fitness and motor competence in children and adolescents with healthy weight: a systematic review and meta-analysis,"

International Journal of Environmental Research and Public Health, Jun. 2021, [Online]. Available: https://doi.org/10.3390/ijerph18136965. [Accessed: Jun. 25, 2024].

[22] C. B. Oliveira et al., "Effects of active video games on children and adolescents: A systematic review with meta-analysis," *Scandinavian Journal of Medicine & Science in Sports*, Jan. 2020, [Online]. Available: https://doi.org/10.1111/sms.13539. [Accessed: Jun. 25, 2024].

[23] I. K. dos Santos et al., "Active Video Games for Improving Mental Health and Physical Fitness—An Alternative for Children and Adolescents during Social Isolation: An Overview," IJERPH, Feb. 2021, [Online]. Available: https://doi.org/10.3390/ijerph18041641. [Accessed: Jun. 25, 2024].

[24] S. Hollister, "The FDA just approved the first prescription video game — it's for kids with ADHD," The Verge, Jun. 16, 2020, [Online]. Available: https://www.theverge.com/2020/6/15/21292267/fda-adhd-video-game-prescription-endeavor-rx-akl-t01-project-evo. [Accessed: Jun. 25, 2024].

[25] M. Kowal, E. Conroy, N. Ramsbottom, T. Smithies, A. Toth, and M. Campbell, "Gaming Your Mental Health: A Narrative Review on Mitigating Symptoms of Depression and Anxiety Using Commercial Video Games," *JMIR Serious Games*, Jun. 2021, [Online]. Available: https://doi.org/10.2196/26575. [Accessed: Jun. 25, 2024].

[26] A. Boldi and A. Rapp, "Commercial video games as a resource for mental health: A systematic literature review," *Behaviour & Information Technology*, Jul. 2021, [Online]. Available: https://doi.org/10.1080/0144929X.2021.1943524. [Accessed: Jun. 25, 2024].

[27] I. Peñuelas-Calvo et al., "Video games for the assessment and treatment of attention-deficit/hyperactivity disorder: a systematic review," *European Child & Adolescent Psychology*, Jan. 2022, [Online]. Available: https://doi.org/10.1007/s00787-020-01557-w. [Accessed: Jun. 25, 2024].

[28] A. Gordillo, D. Lopez-Fernandez, and E. Tovar, "Comparing the effectiveness of video-based learning and game-based learning using teacher-authored video games for online software engineering education," *IEEE Transactions on Education*, Jan. 25, 2022, [Online]. Available: https://doi.org/10.1109/TE.2022.3142688. [Accessed: Jun. 25, 2024].

[29] L. Martinez, M. Gimenes, and E. Lambert, "Entertainment Video Games for Academic Learning: A Systematic Review," *Journal of Educational Computing Research*, Jan. 2022, [Online]. Available: https://doi.org/10.1177/07356331211053848. [Accessed: Jun. 25, 2024].

[30] D. R. Ross, D. H. Finestone, & G. K. Lavin, "Space Invaders obsession," *JAMA*, Sep. 10, 1982. [Online]. Available: https://doi.org/10.1001/jama.1982.03330100017009. [Accessed: Jun. 25, 2024].

[31] J. Clement, "Weekly time spent playing video games according to adults in the United States as of December 2023, by age group," Statista, Feb. 2024. [Online] Available: https://www.statista.com/statistics/202839/time-spent-playing-games-by-social-gamers-in-the-us/. [Accessed: Jun. 25, 2024].

[32] C. Tilds, "Men and Women Ages 18-25 & 26-35 Spend More Time Playing Video Games Than Watching Broadcast TV," Frameplay, Mar. 2021, [Online]. Available: https://frameplay.com/resource/men-and-women-ages-18-25-26-35-spend-more-time-playing-video-games-than-watching-broadcast-tv/. [Accessed: Jun. 25, 2024].

[33] J. Sherer, "Internet Gaming," American Psychiatry Association, Jan. 2023, [Online]. Available: https://www.psychiatry.org/patients-families/internet-gaming. [Accessed: Jun. 25, 2024].

[34] Y. X. Gao, J. Y. Wang, and G. H. Dong, "The prevalence and possible risk factors of internet gaming disorder among adolescents and young adults: Systematic reviews and meta-analyses," *Journal of Psychiatric Research*, Oct. 2022, [Online] Available: https://doi.org/10.1016/j.jpsychires.2022.06.049. [Accessed: Jun. 25, 2024].

[35] P. Grinspoon, "The health effects of too much gaming," Harvard Health Publishing, Dec. 22, 2020. [Online]. Available: https://www.health.harvard.edu/blog/the-health-effects-of-too-much-gaming-2020122221645. [Accessed: Jun. 25, 2024].

[36] "Hate is No Game: Hate and Harassment in Online Games 2023," Anti-Defamation League, Feb. 6, 2024. [Online]. Available: https://www.adl.org/resources/report/hate-no-game-hate-and-harassment-online-games-2023. [Accessed: Jun. 25, 2024].

[37] J. Howarth, "How Many Gamers Are There? (New 2024 Statistics)," Exploding Topics, Jun. 11, 2024, [Online]. Available: https://explodingtopics.com/blog/number-of-gamers. [Accessed: Jun. 25, 2024].

[38] C. Figueiredo, "Five Things Parents Should Know About Content Moderation in Video Games," Family Online Safety Institute, Nov. 13, 2018. [Online]. Available: https://www. fosi.org/good-digital-parenting/five-things-parents-should-know-about-content-moderation-video-games. [Accessed: Jun. 25, 2024].

[39] M. Teo, "Nurturing the Wellbeing of Content Moderators in Gaming," Zevo Health, Feb. 26, 2024. [Online]. Available: https://www.zevohealth.com/blog/nurturing-the-wellbeing-of-content-moderators-in-gaming/. [Accessed: Jun. 25, 2024].

[40] Archimtiros, "Malicious WeakAura Replaces Auction House Purchases with Overpriced Scams," Wowhead TBC, May 28, 2021. [Online]. Available: https://tbc.wowhead.com/news/malicious-weakaura-replaces-auction-house-purchases-with-overpriced-scams-322567. [Accessed: Jun. 25, 2024].

[41] L. Grustniy, "A trick auction to steal gold in World of Warcraft," Kaspersky Daily, Jun. 16, 2021. [Online]. Available: https://www.kaspersky.com/blog/wow-weakauras-auction-scam/40280/. [Accessed: Jun. 25, 2024].

[42] Dhivya, "Tweaks Stealer Attacks Online Game Users Abusing YouTube & Discord," Cyber Security News, Mar. 13, 2024. [Online]. Available: https://cybersecuritynews.com/tweaks-stealer-attacks-game/. [Accessed: Jun. 25, 2024].

[43] T. Meskauskas, "How to eliminate Tweaks stealer from infected computers," PC Risk, Apr. 2, 2024. [Online]. Available: https://www.pcrisk.com/removal-guides/29351-tweaks-stealer. [Accessed: Jun. 25, 2024].

[44] "Gamers beware: The risks of Real Money Trading (RMT) explained," Malwarebytes Labs, Sep. 10, 2021. [Online]. Available: https://www.malwarebytes.com/blog/news/2021/09/gamers-beware-the-risks-of-real-money-trading-rmt-explained. [Accessed: Jun. 25, 2024].

[45] "Trading Items and Services for Real Money," Battle.net, May 2024. [Online]. Available: https://us.battle.net/support/en/article/269874. [Accessed: Jun. 25, 2024].

[46] Merriam-Webster, "gambling - noun," Merriam-Webster. [Online]. Available: https://www.merriam-webster.com/dictionary/gambling. [Accessed: Jun. 25, 2024].

[47] Taylor & Francis Group, "Purchasing loot boxes in video games associated with problem gambling risk, says study," Science Daily, Dec. 22, 2022. [Online]. Available: https://www.sciencedaily.com/releases/2022/12/221202112513.htm. [Accessed: Jun. 25, 2024].

[48] D. Zendle, R. Meyer, and H. Over, "Adolescents and loot boxes: links with problem gambling and motivations for purchase," *Royal Society Open Science*, Jun. 2019. [Online] Available: http://dx.doi.org/10.1098/rsos.190049. [Accessed: Jun. 25, 2024].

[49] S. G. Spicer, C. Fullwood, J. Close, L. L. Nicklin, J. Lloyd, and H. Lloyd, "Loot boxes and problem gambling: Investigating the 'gateway hypothesis,'" Addictive Behaviors, Aug. 2022, [Online]. Available: https://doi.org/10.1016/j.addbeh.2022.107327. [Accessed: Jun. 25, 2024].

[50] P. J. Etchells, A. L. Morgan, and D. S. Quintana, "Loot box spending is associated with problem gambling but not mental wellbeing," *Royal Society Open Science*, Aug. 17, 2022, [Online]. Available: https://doi.org/10.1098/rsos.220111. [Accessed: Jun. 25, 2024].

[51] C. Bedford, "11-year-old boy spent £464 on game microtransactions and loot boxes," Leicestershire Live, Dec. 5, 2022, [Online]. Available: https://www.leicestermercury.co.uk/news/local-news/11-year-old-boy-spent-7877406. [Accessed: Jun. 25, 2024].

[52] Trends Desk, "13-year-old spends Dh235,071 on online gaming in China, nearly empties mother's bank account," Khaleej Times, Jun. 14, 2023, [Online]. Available: https://www.khaleejtimes.com/offbeat/13-year-old-spends-dh235071-on-online-gaming-in-china-nearly-empties-mothers-bank-account. [Accessed: Jun. 25, 2024].

[53] R. Thubron, "A 13-year-old spent $64,000 of her parents' money on mobile games without them realizing," Techspot, Jun. 7, 2023, [Online]. Available: https://www.techspot.com/news/98980-13-year-old-spent-64000-parents-money-mobile.html. [Accessed: Jun. 25, 2024].

[54] E. Killham, "Japan's Consumer Affairs Agency declares 'complete gacha' illegal," Venture Beat, May 18, 2012. [Online]. Available: https://venturebeat.com/games/complete-gacha-illegal/. [Accessed: Jun. 25, 2024].

[55] N. Straub, "Every Country With Laws Against Loot Boxes (& What The Rules Are)," Screen Rant, Oct. 5, 2020, [Online]. Available: https://screenrant.com/lootbox-gambling-microtransactions-illegal-japan-china-belgium-netherlands/. [Accessed: Jun. 25, 2024].

[56] J. Ye, "China announces rules to reduce spending on video games," Reuters, Dec. 22, 2023, [Online]. Available: https://www.reuters.com/world/china/china-issues-draft-rules-online-game-management-2023-12-22/. [Accessed: Jun. 25, 2024].

[57] T. Gerken, "Video game loot boxes declared illegal under Belgium gambling laws," BBC News, Apr. 26, 2018, [Online]. Available: https://www.bbc.com/news/technology-43906306. [Accessed: Jun. 25, 2024].

[58] Department for Digital, Culture, Media & Sport, "Government response to the call for evidence on loot boxes in video games," Gov.UK, [Online]. Available: https://www.gov.uk/government/calls-for-evidence/loot-boxes-in-video-games-call-for-evidence/outcome/government-response-to-the-call-for-evidence-on-loot-boxes-in-video-games. [Accessed: Jun. 25, 2024].

9 From Swings to Smartphones – The Impact of Social Media on Kids

Sometimes the biggest challenge parents face with raising kids is what other parents do. When my own daughter was 11 years old, she kept asking for a smartphone. Almost every parent is naturally apprehensive when it comes to giving their child(ren) a smartphone because they know what it can do, what it can access, and the door it opens into the online world. Parents don't automatically think 'ah now they can access the Encyclopedia Britannica, read about Einstein's theory of special relativity and all will be good'. Being completely ok with giving your kid a smartphone should be on par with being open to your eleven-year-old taking up smoking.

You are an adult, and you know what's lurking on the internet; the porn, the sexualization of minors, the trolling, the doxing, the cyberbullying, the catfishing, and even the pedophiles impersonating other kids. Social media's version of the horsemen of the apocalypse for your child's psychological development. Then there are the ones that parents don't often think of: the viral dares, dangerous challenges, the fads and extreme dieting. I kept telling my daughter not now but maybe for your next birthday. But the response she gave was the one that parents dread the most. She went on to list the names of all the other kids

DOI: 10.1201/9781032679389-10

in her class that had been given one by their parents. Her best friend Molly even had one. And now as a parent you are boxed in big time. You say you don't care about what other parents do (but you do and can't admit it to them) and at the same time you know that the peer status of your kid is devaluing as rapidly as a third-world country's currency. Suddenly, your kid is the odd one out and they are feeling it. Other parents have fallen like dominos around you, and eventually you give in too. Maybe Molly was given the phone in the first place so her parents could call her if they were going to be late picking her up. Maybe there were other valid reasons, but Molly's parents have set in motion a cascade of events that end up with you as a parent in a fairly unpleasant and unwelcome position. You want to protect your child but can't actually tell them the actual reasons you are resisting giving them a wormhole to the internet. As every parent knows, in this situation logic with your kid does not prevail and it's akin to an irresistible force meeting an unmovable object. You rationalize the benefits (which do exist), you say you will limit screen time, install safety controls, monitor what they are doing (which you do) and still in the back of your mind is that this move has the potential to explode in your face. You hope and pray every day the bomb doesn't go off. Every parent has heard the horror stories of cyberbullying and kids committing suicide as a result. Parents are struggling and even end up writing anonymous letters to pediatricians asking for help like the following:

> Dear Pediatrician. My middle schooler really wants a smartphone, but I'm not so sure. He says that most of the kids in his class already have a phone, and he feels left out. I'm worried about him spending too much time on the phone. Plus, I've heard scary stories about kids sending inappropriate messages to one another. Is there a best age to give your child a smartphone?

The advent of the smartphone and the rise of social media has changed kids. Kids today are born into a fully digital world unlike all previous generations, and some have argued that the world shifted from a play-based childhood to a device-based one in 2010 with the advent of front-facing cameras on phones [1]. These phones launched the world of the 'selfie' along with the platforms where they could be posted for the world to see. Look around and you can see how even very young children use the internet. I've seen two-year olds work their way around an iPad at the speed of a computer programmer to find what they want to watch but luckily their interests are limited to just watching cartoons. That's a basic reinforcement relationship from a very young age. Click-look-laugh-repeat. As we will see later that basic relationship of click-look-reinforce has been taken to a stratospheric level by social media platforms. But for now, a quick example of the over-all meta-effect of the effect of social media on kids just involves considering the following basic rule of thumb: what kids want to be when they grow up is what they find cool as a kid.

In the 1960s and 1970s the number one thing kids wanted to be when they grew up were astronauts. Today, one of the top five things kids consistently want to be when they grow up is to be an 'influencer' on one or another social media platform and that desire can only come directly from social media platforms [2]. For boys, it's number two after being a professional athlete whereas girls still appear to be somewhat sensible, and for them, it comes at number five. I imagine that's not the profession a lot of parents would like their children to gravitate toward but maybe the kids aren't entirely crazy. The term influencer is basically a fancy word for a salesperson trying to flog stuff on social media [3]. However, as you might expect, influencers themselves define it more 'influentially'. To them, it's the power to affect the purchasing decisions of others because of his or her authority, knowledge, position, or relationship with his or her audience [4]. That is quite the skill set for someone to have and not one that many parents would reject. And maybe the kids aren't wrong, because according to some internet sources (which requires a pinch of salt) like Influencer Marketing Hub [5], the market size for influencer marketing has doubled since 2019 and was valued at $16.4 billion worldwide in 2022. It seems the kids are following the money.

THE SURGEON GENERAL RINGS THE ALARM BELLS

The question as to whether social media has a negative effect on the mental health of young people is a complicated one. Today's kids do not know a world without digital technology but it's fair to say that the digital world was not built at the outset with concerns over their healthy mental and physical development in mind [6]. The owners of the big social media platforms never started from this position and have been playing a combination of catch-up and dodgeball ever since. The alarm bells around the impact of social media on kids have been getting a lot louder of late and to such an extent that in 2021 US Surgeon General Dr. Vivek Murthy issued a national report where he indicated that the use of social media platforms was one of the key drivers of a mental health crisis in this age group [7]. That is a fairly serious claim to make and one that should raise some concern and might have you reaching toward the router in your home. Nor does it appear to be a case of 'oops we never thought about that' among the tech giants when you consider that in 2023 the New York District Attorney Letitia James, along with top prosecutors from more than 30 other states in the US, filed suit against the social media giant Meta (the owner for Facebook and Instagram). The suit alleged that the company deliberately put features in place on Instagram and Facebook to intentionally 'addict' young people for profit [8]. Basically, the accusation is that many of the social media platforms are intentionally designed in a way that ramps up habitual use by teenagers and exposes them to content that, to put it mildly, is often not in their best interest. It's not a million miles away from a drug pusher hanging around a school with free doggie bags to capture a future clientele.

In early 2024, the chief executives of Meta, X, and TikTok and others were summoned to testify before US Congress over alleged harm to children's mental health over the use of social media. Promises to make platforms safer for teenagers have been made. Not for the first time either [9]. The efforts made by some platforms to limit the time kids spend on them have also proven to be somewhat half-hearted. For instance, in 2023 TikTok introduced a one-hour time limit for those under the age of 18, but once that limit is reached users can simply enter a password and keep scrolling. Irrespective of the promises (and the efforts) to make social media platforms safer, there is a growing movement calling for a complete ban on the use of social media platforms for kids under the age of 16 [10].

But before we get into talking about the harm that social media can potentially do to children, it has to be said that it is not all bad which is exactly what makes debates about the impact of social media somewhat complicated in the first place. For instance, adolescents frequently say social media is a forum where they can express their creative side and where they find positive identity-confirming content [11]. Yet, it can also help with aspects of mental health as female teenagers with mild symptoms of depression say their life would be worse without access to social media. They also say that social media content can help them understand their experiences and feelings. Among the benefits, over half of adolescents report that social media helps them feel more accepted, two-thirds say they can easily access others who can support them through tough times, and over three-quarters say they feel connected to what's going on in their friends' lives. This is all good, right? Well, yes but if you look at the underlying theme of these three benefits you might also notice the very thing that connects them together and which can make children and young teenagers more vulnerable to harm from social media usage in the first place. At a very simple level, the need to belong to their peer group becomes dominant in the early teen years, and conversely so is the sensitivity to social comparison, social validation and social rejection. That's something we all know, and we have all been there but what most of us don't know are the changes going on in the brain which underpin why teenagers are more vulnerable to the negative consequences of social media usage than adults.

WHY KIDS ARE MORE VULNERABLE TO SOCIAL MEDIA

There are reasons we don't let kids vote, drive cars, drink alcohol and own guns. It's simply because they are not psychologically developed enough across a whole load of areas like emotions, complex thinking, a solid sense of right and wrong and, perhaps most crucially of all, is not meaningfully appreciating risk [12]. No pun intended but it's no accident that teenagers are particularly susceptible to accidents. They are not very good at calculating risks when doing things like heading off to climb the tallest tree in the neighborhood, jumping off the rooftop of their home and skateboarding without a helmet. This is not being negative about kids, that's just the way they are. Their brains are simply still developing.

To understand why the Surgeon General is getting in a bit of a flap about social media and children's mental health, it's worthwhile to look at a little bit of science. Starting around age 10, kid's brains undergo a major change that pushes them to seek social rewards, including attention and approval from their peers [13]. That stuff is gold dust for teenagers but it's also the age where kids begin to engage in social comparison with others. It's also the start of when your kid begins to think that hanging out with their parents is the epitome of uncool. This is because at around this age, receptors for the 'happy hormones' oxytocin and dopamine have begun to multiply in a part of the brain called the ventral striatum which has the effect of making preteens extra sensitive to social things like attention and admiration from others (except their parents, aunts, uncles and anyone they consider ancient). Sitting right next door to the ventral striatum in the brain is another area called the ventral pallidum and this area is involved in motivation, so one area makes you feel happy and the other motivates you to seek out what makes you happy. But the brain itself does not care where this social attention comes from and will pump out happy hormones just as easily from a 'like' on social media to someone complimenting you in person about how you look; yet, because it's a brain reward center you will be motivated to seek it out more.

The reason why kids are more vulnerable to this type of reward is that adults tend to have a fixed sense of self that relies less on feedback from their peers. Added to that is the fact that in adults, another area of the brain called the prefrontal cortex, is more developed which in turn regulates emotional responses to social rewards and social rejection. In other words, as an adult you're not going to fall apart when someone clicks the thumbs-down button on your post, and you will probably brush it off fairly easily. Yet, some like the Surgeon General believe that herein lies a potential recipe for disaster for teenagers, a massive shift toward social media for needed social contact but with the brain regions involved in self-control and emotional regulation not fully developed [14]. It might seem here that the argument I'm making is that kids' brains are just not developed and that's what makes them vulnerable to social media usage but it's not that simple. As we will get to later, it also seems that social media platforms may well be exploiting this vulnerability or at the least, not taking it into consideration enough.

SMART PHONE AND SOCIAL MEDIA USE AMONG KIDS

Few adults can get by without a smartphone and kids grow up to become adults so what is the right age for a kid to get their hands on one? It's a great question but given the current situation it is also a rather pointless one. In the USA, almost half of ten-year olds have a smartphone and by the age of 14, the horse has well and truly bolted from the stable with that figure climbing to 91% [15]. Social media use is pretty much universal among teenage girls with a whopping 98% of girls surveyed and not a whole lot lower for boys [16]. Whatever the answer is to the question of age, the genie is out of the bottle, and it is not going back in on its own. Teachers especially are sick to death of younger and younger and more and more kids having phones in schools. Schools have become ground-zero for the epidemic of smartphones among teenagers especially. The problem of kids spending so much time on their phones in school, combined with parental worries about social media, has led to some serious soul-searching. Some schools have tried the approach of only allowing old-style brick phones to enter the premises, but this methadone-type approach to phone substitution has not been very successful. As of right now, almost three-quarters of primary schools in the United Kingdom either collect phones from kids on arriving at school or have introduced 'lockboxes' where the smartphones are now experiencing COVID-19 levels of lockdown [17]. Some primary schools in Ireland have banned phones in any shape or form completely. Luckily, kids are not very good at protesting or starting revolutions and for the most part, just get on with it. But parents are concerned (at least they should be), and that concern only seems to be growing. When asked, almost half of American parents believe their child or children's mental health has deteriorated in the previous year because of social media use [18] and an even larger number believe that high social media usage has rendered their offspring unable to socialize properly [19]. Parents are at the coalface of social media usage among their kids and are begging for people to listen.

SOCIAL MEDIA AND TIME-THIEVES

The word addiction has crept into the conversation (and parents' nightmares) about teenagers' usage of social media platforms in a rather big way. The media is awash with statements like '1 in 5 U.S. parents worry their teen is addicted to the internet' [20] and 'social media addiction is real! 40% Indian parents admit their teen kids addicted to smartphones' [21]. The word addiction conjures up images for parents of their kids being zoned out, not leaving the bedroom, falling behind in school, and a raft of other fears. The one thing that is making parents scream addiction is the amount of time kids are spending on social media platforms and it's not hard to see why. Some surveys report that the average amount of time a teenager is glued to a social media platform is 4.8 hours *per day* [22]. Girls it seems are the worst offenders and their average time is 5.5 hours with boys spending just under an hour less per day. The chances are if you're a parent and your kid is back from school, pretty much every time you look at what they are doing, their face is not far from a screen. Almost five hours doesn't leave much time for the other things teenagers need to be doing like homework, family time, and hanging out with friends in an actual non-virtual place. It's no wonder parents are seriously worried especially when information like the following is thrown into the mix: teenagers who spend more than three hours per day on social media have double the risk of experiencing poor mental health outcomes, such as symptoms of depression and anxiety [23]. No wonder the collective sound of parents screaming 'get off the phone' is booming across the country.

The three biggest social media time-thieves are YouTube, TikTok, and Instagram accounting for almost 90% of teenager social media time [23]. TikTok especially seems to be the equivalent of a digital opiate for kids with some studies reporting a massive average time of 159 minutes per day on that single platform alone. Without getting all technical on what addiction in the context of social media means, almost half of girls asked about their use of TikTok say they feel addicted to the platform and use it more than they intended. Why would this be? An answer to that requires a little exploration of how social media platforms like TikTok are designed in the first place and when you understand that, it becomes easier to understand the massive amounts of time being spent on it In the 1980s there were books on how to give up smoking, and it may not be long before we see the equivalent for an easy way for trying to give up platforms like TikTok.

TikTok is explicitly designed (but so are other social media platforms) to keep you on the hamster wheel of scrolling. It's desperate for your attention because greater user engagement increases advertising revenue (just follow #tiktokmademebuyit). That's the bottom line and in fact, it generates a lot of it. In 2023, the company that owns TikTok reported a revenue of 16 billion dollars in the USA alone [24]. The platform uses something called algorithmic targeting and this algorithm has been described as the equivalent of a 'superpower' [25]. It creates what is called the 'for you page (FYP)'. Basically, TikTok uses artificial intelligence (AI) and machine learning to analyze the videos you watch, the amount of time you spend on a video, what you like, and what you share and comment on. Every one of these little actions carries a certain amount of weight and signals back to the platform how much you as the user engaged with the content. So, each and every time a kid watches a video, TikTok is collecting information on them and in just a few hours, the algorithm can detect their musical tastes, their hobbies and interests, what they find funny and even if they are depressed [26]. The algorithm knows what the kid likes, then blends that familiar content with new content to keep it fresh and, what you might call, addictive. It's not magic. It's just a clever formula designed to evolve to keep the user engaged. It's simply giving you what you like to watch, and you'll keep watching because it's what you like. Essentially, the more videos that show up on your feed that you relate to, the more likely you are to stay on the platform or return to it. TikTok relies heavily on how much time you spend watching each video to steer you toward more videos that will keep the user scrolling, and that process can sometimes lead young viewers down some dangerous rabbit holes. It can steer them toward content that promotes body image problems, thoughts of suicide or even self-harm. There is also an unfortunate and potentially damaging trend of kids using social media to diagnose mental health conditions in themselves after watching influencers. Up to half of teenagers have ended up giving themselves some diagnostic label [27].

THE PROBLEM OF ALGORITHMS

This is at the heart of the problem with algorithms like that used by TikTok because it serves up on the screen what you engage with and if depressive content keeps you engaged then that's what you will get. There are far too many heart-breaking stories of parents whose teenage children have committed suicide and who later find a steady stream of videos about depression, break-ups, death and suicide on the social media platforms used by their children prior to their death. The majority of these parents believe that their children were fed a diet of negativity by the algorithm, and some have moved to take legal action against a platform as a result [28].

Only a few years ago, the *Wall Street Journal* published an investigation that involved monitoring more than 100 automated accounts to see how TikTok's algorithm works. Within just over half an hour the newspaper reported, a bot that was programmed to engage with videos about depression was then fed a stream of content that was over 90% about sad topics [29]. Hours of exposure to that kind of content is not going to have a neutral effect on a teenager who might already be struggling. At the time, TikTok representatives got a little defensive and claimed that the investigation using bots wasn't representative of human behavior but an argument like that one is not going to win any public relations battle with parents. TikTok might be waking up. In mid-2024, the company acknowledged that certain types of content might be 'problematic if viewed in clusters. This includes content such as dieting, extreme fitness, sexual suggestiveness and sadness' [30]. Yet at the same time they might be hedging a little when they also state that this kind of information *may be* eligible for the FYF, but they will interrupt repetitive content patterns to ensure it is not seen *too often*. While that's all sounding positive, what they don't say clearly, at all, is how do you define too often? What's the algorithm for that exactly? It's also worth remembering that TikTok is almost eight years old having been released in September 2016. It seems to have taken them quite a long time to get to this point and perhaps with a little bit of prompting from parental lawsuits and congressional maulings.

Report after report in some manner recommends that social media platforms like TikTok, Instagram and Snapchat should be designed in ways that promote well-being and reduce exposure to hazards. Except that isn't good for business. I admit I use Instagram and one thing I noticed quickly that what's appearing on the screen is certainly not random; click a like on one motorbike feed and for the next two days it is motorbikes all over the place and then it starts throwing up pictures of scantily clad females. I'm guessing there is an algorithm at play linking the two (maybe it's a biker thing). That's all fun and games until you are a teenage girl with high level of symptoms of depression already because here is a not-fun fact: Girls with moderate to severe depressive symptoms are nearly three times as likely as girls without depressive symptoms to come across harmful suicide-related content on social media platforms at least monthly. But what makes it more complicated is they also report regularly coming across helpful mental health information and resources on social media. It's a mix of the good, the bad and, as we will get to, the idealized beauty standard.

WHAT ARE KIDS EXPOSED TO ON SOCIAL MEDIA

Parents naturally exercise a healthy dose of paranoia when it comes to their kids and social media. The bad news is that's it largely warranted. A look at some of the things happening to kids on Instagram and Snapchat will make you weak at the knees fairly quickly. More than half of girls aged between 11 and 15 years old have been contacted by strangers in a way that made them feel uncomfortable [31]. Rather than going into a definition of the word uncomfortable, take a moment and think about what happened to a girl called Clare. She made a video of herself dancing and posted it on TikTok. That's something a lot of children do on a daily basis. Clare's video got a lot of attention and almost 5,000 likes but a lot of these were sexualized comments. However, Clare was only 11 years old at the time [32]. Some of the more specific statistics are scary. Overall, almost 50% of girls say they have been at the receiving end of unsolicited messages on a social media platform with Instagram (58%), Snapchat (57%) and TikTok (46%) being the main platforms where this behavior takes place.

The problem of exposing teenagers to sexualized content on social media isn't just confined to seedy individuals scrolling and making lewd comments. Most influencers on social media platforms are in fact women [33]. However, the body seems to play a big role in influencers' selfies, and they can spend an enormous amount of time and effort generating the perfect shot in order to attract attention and therefore increase the chances of making money. An Instagram post might grab your attention for the hot, young, scantily clad girl posing suggestively on a kitchen counter, but hey, turns out she was just peddling amino acids all along [34].

Unfortunately, and sadly, one of the easiest ways for female influencers to attract attention on social media is through a sexualized aesthetic. For some proof of this, one study took a random group of female influencers on Instagram and analyzed the images, interactions, and comments of the influencers themselves over a period of several months. It found a continuum of pornified self-representations ranging from what the authors of the study called 'softer' references where influencers adopted poses to highlight sexualized body parts. They also tended to employ 'porn chic' gestures such as gently pulling their hair, touching their parted lips and even squatting with legs spread to the camera. There were even instances of images that were difficult to differentiate from mainstream commercial pornography [35]. So, in effect, the sexualized material that teenagers are exposed to can come from several directions on social media platforms.

Then, of course, there are the effects of social media on body image among teenagers. Have you ever heard of something called 'Snapchat dysphoria'? If you are a plastic surgeon, you probably already have. It's a truism that people like to post the best images of themselves on social media and that applies to teenagers just as much as adults. Groucho Marx once quipped 'these are my principles, and if you don't like them…. well, I have some others'. Well, Instagram and Snapchat have applied pretty much the same concept but this time to images of yourself. If you don't like this image of yourself, we will give you another and better one. Both of these platforms now provide a whole range of image-altering filters that allow users to change their skin tone, soften fine lines and wrinkles, alter the size of their eyes, lips, and cheeks, and change various aspects of their physical appearance [36]. A few tuned-in plastic surgeons in the USA, however, noticed that the physical alterations that some people were requesting for themselves bore more than a passing resemblance to the images produced by social media filters. Hence the term Snapchat dysphoria. Thankfully, some gentle guidance toward the need for counseling was provided. Granted this is taking filtering to the extreme but it does make you ask questions about the effects of social media on how young people are feeling about themselves in terms of body image.

The question surrounding the role of social media becomes even more important when you realize that adolescence is a vulnerable period for the development of body image issues, eating disorders and mental health difficulties. A teenager spending three hours a day on a social media platform is being exposed to hundreds if not thousands of already filtered images of peers, influencers, fashion models and celebrities which can lead to an internalization of beauty ideals that are unattainable for pretty much everyone [37]. Appearance comparisons inevitably occur and dissatisfaction with how you view yourself being the most likely outcome. While all genders are susceptible to these appearance comparisons, it appears to be the case that female teenagers are the most susceptible to pressure from the unrealistic standards of beauty portrayed on social media platforms and the frequent endorsement of the 'thin-ideal'. This is because first of all, pubertal maturation on its own is often a threat to female adolescent's body esteem due to unwanted bodily changes such as increased body fat. Second, there is also a vicious cycle whereby teenagers rely on signs of appearance-focused popularity to meet their need for peer acceptance and social status which results in relentless physical comparison [38].

Up to now, a lot of studies have looked at the effect of social media on adolescent body image and study after study reaches the same general conclusion; excessive social media use leads to unhealthy body esteem because of these internalized standards, which aggravates appearance comparisons and anxiety regarding negative appearance evaluation [39]. When you ask adolescents simply and directly about the impact of social media on their body image, almost half of those aged 13–17 say social media makes them feel worse, and only a small percentage say it makes them feel better.

Skeptics of social media at this point might well point to a chicken and egg and egg scenario. Is it the case that kids who already have body image and mental health issues spend more time on social media or is it that social media use leads to greater body image and mental health issues? Before you think the jury is out on that one, one interesting study points a finger at social media in at least sustaining poor body image in teenagers. Researchers took a group of young adults between 17 and 25 years of age who already spent two hours per day on social media and who had symptoms of anxiety and depression. For a three-week period, they reduced their social media usage by half and compared them to a group who did not reduce their time on the platform. Unsurprisingly, the 'reducers' reported that they felt happier with their appearance and their body weight as a result [40].

Ask any adult what the other kids in their school were like when they were growing up and I'll bet that almost everyone will recount that some other kids were just bullies and made their life miserable. We should remember that kids are not paragons of virtue and innocence, and some of them can be downright mean and nasty to each other. And some stay that way even as the years accumulate, and they graduate to become seasoned online trolls. Nastiness can start early. Before the rise of social media, you could at least go home after school and escape from your school or neighborhood bully for a while. These days they can follow you into your home. There is little chance of escape. These creeps follow you into the digital world and have given rise to what we now call 'cyberbullying'. While your average bully is a bit of a coward and tends to avoid face-to-face confrontations, the digital world allows bullies to now thrive at their trade. It suits them perfectly. Sometimes you might never even know who the bully is because these days they can create a fake profile and launch their harassment missiles at you in complete anonymity. But what makes it far worse is that when stuff gets posted that's designed to harass you, it can spread like wildfire and worse, it can stay accessible for a very long time.

The second leading cause of death among teenagers is the USA is suicide. That is not a pleasant statistic for the leader of the free world, and some have claimed that social media platforms are contributing to this. Cyberbullying has been associated with the suicide of victims in far too many cases, a phenomenon that has generated the term cyberbullicide [41]. The effects of cyberbullying are quite frankly awful. Take a moment and reflect on the following statistics. A kid who is the victim of cyberbullying is more than four times as likely to report thoughts of suicide and attempts as a result [42] and a third of those cyberbullied go on to develop social anxiety which can often travel well into adulthood [43]. Almost half of all kids in the USA between 13 and 17 years old report that they have been cyberbullied [44]. Cyberbullying on social media platforms comes in several shapes and sizes: offensive name-calling, exclusion, spreading false rumors, receiving explicit images they did not request, constantly being asked what they are doing, physical threats, and sharing explicit images of them that they did not consent to. Girls are about twice as likely as boys to be victims and perpetrators of cyberbullying. It's interesting what teens themselves say about what would help stop cyberbullying. Half of them think that cyberbullies should have a little visit from law enforcement, and four in ten think that social media companies should look for and delete posts that reflect bullying and an almost equal number think that social media companies should force people to use their real names when posting on social media.

HOW SOCIAL MEDIA AFFECTS SLEEP

If you ask parents their top concerns about kid's use of social media platforms, they include becoming addicted to social media, receiving predatory messages, receiving or engaging in X-rated content or conversations, and talking to strangers [45]. These are the worries that keep a lot of parents up at night but while the parents themselves are awake, far too often so are their kids and social media is often the culprit. Now you might shrug that off and say well they are just going to be a little tired for the next day and no major harm done except the risk of harm is quite real. Earlier we talked about how the brain in children undergoes changes from around the age of 10 but what is sometimes forgotten is an appreciation of how critical sleep is to overall brain development in children. Time to buckle up, take a deep breath and reflect on the following shopping list of negative health outcomes associated with routine sleep deprivation in children and adolescents: obesity, increased cholesterol, diabetes, poor memory and attention, decreased IQ and academic performance, and increased depression, anxiety, aggression and ADHD [46]. Sleep is incredibly important for both physical and mental health in kids and the statistics don't make for pleasant reading. Up to one-third of infants and school-age children have poor sleep health and over three-quarters of high school students sleep less than the recommended eight hours per night [47]. There are of course a multitude of reasons for this epidemic of sleep-deprived daytime zombie kids, but social media usage does seem to be one of them. In fact, there are three factors associated with social media that are drivers of poor sleep.

First, the light that is emitted from handheld devices like smartphones, even with a night filter or a blue light filter is enough to suppress levels of melatonin which is the main hormone that signals to the body the onset of sleep [48]. Second, what is being watched is a problem. Fast-paced imagery like you see on TikTok highly stimulates the brain while awake making getting to sleep more difficult. Added to that is that when kids do get to sleep, the brain may not sustain deep nondream sleep because the brain is still processing those images [49]. There is also the simple fact that if the child is looking at social media because of reasons like FOMO, they are simply awake and the time available for sleep is evaporating and remember, like rust, social media platform algorithms don't sleep. They keep on working.

QUICK WRAP UP...

At the end of this, it's up to you to make your own conclusions about the impact that social media is having on large numbers of kids. Platforms like Instagram and TikTok have been highlighted by others as the ones that may be the most damaging to the mental health of young people. For example, the whistleblower Frances Haugen reported in 2021 that Facebook had conducted its own research and found that it is not just that Instagram is dangerous for teenagers, but that it is distinctly worse than other forms of social media [50]. Whistleblowers aside, there are a lot of scientific papers examining the effect of social media and mental health in kids and it does seem to be the case that the risks are legitimate and outweigh the benefits for many. At the outset of this chapter, the concerns of the US Surgeon General about the impact of social media were highlighted. In May 2024, he took his concerns a step further when called on the US Congress to require labels on social media sites to remind users that the platforms had 'not been proved safe' and were 'associated with significant mental health harm for adolescents' [51]. It's unlikely he is doing this just to keep himself busy. While the genie may not go back in the bottle, we should certainly consider whether access to social media might be wiser after the age of 16. Maybe it should be a case of better safe than sorry. Without a doubt, that would be resisted, and I have little doubt that the owners and shareholders of the big social media platforms would conjure up lots of positive findings to counter the negative.

REFERENCES

[1] J. Jargon. "Jonathan Haidt blamed tech for teen anxiety. Managing the blowback has become a full time job," The Wall Street Journal, May 10. 2024. [Online]. Available: https://www.wsj.com/tech/personal-tech/jonathan-haidt-anxious-generation-book-smartphones-676bcadb?reflink=desktopwebshare_permalink [Accessed: Jun. 22, 2024].

[2] F. Zandt. "Most popular future jobs with United States teenagers," Statista, Oct. 11, 2023. [Online]. Available: https://www.statista.com/chart/31014/most-popular-future-jobs-with-united-states-teenagers/. [Accessed: Jun. 22, 2024].

[3] "Influencer," Cambridge Dictionary, [Online]. Available: https://dictionary.cambridge.org/dictionary/english/influencer. [Accessed: Jun. 22, 2024].

[4] W. Geyser. "What is an influencer?" Influencer Marketing Hub, Feb. 14, 2024. [Online]. Available: https://influencermarketinghub.com/what-is-an-influencer/. [Accessed: Jun. 22, 2024].

[5] Influencer Marketing Hub, Jun. 22. 2024 [Online]. Available: https://influencermarketinghub.com/. [Accessed: Jun. 22, 2024].

[6] J. Haidt. "End the phone-based childhood now". May 13, 2024. [Online]. Available: https://www.theatlantic.com/technology/archive/2024/03/teen-childhood-smartphone-use-mental-health-effects/677722/. [Accessed: Jun. 22, 2024].

[7] "Surgeon General issues new advisory about effects social media use has on youth mental health," U.S. Department of Health and Human Services, May 23, 2023. [Online]. Available: https://www.hhs.gov/about/news/2023/05/23/surgeon-general-issues-new-advisory-about-effects-social-media-use-has-youth-mental-health.html [Accessed: Jun. 22, 2024].

[8] "If your child is addicted to TikTok, this may be the cure," The New York Times, Nov. 17, 2023. [Online]. Available: https://www.nytimes.com/2023/11/17/nyregion/tiktok-social-media-children-addiction.html. [Accessed: Jun. 22, 2024].

[9] P. Ryan. "Addiction to social media is changing perceptions," The National, Feb. 2, 2024. [Online]. Available: https://www.thenationalnews.com/uae/2024/02/02/like-smoking-near-children-addiction-is-changing-how-people-see-social-media-experts-say/. [Accessed: Jun. 22, 2024].

[10] J. Haidt (2024). *The anxious generation: how the great rewiring of childhood is causing an epidemic of mental illness.* Penguin Press.

[11] M. Anderson and J. Jiang, "Teens and their experiences on social media," Nov. 28, 2018. [Online]. Available: https://www.pewresearch.org/internet/2018/11/28/teens-and-their-experiences-on-social-media/. [Accessed: Jun. 22, 2024].

[12] E. Bozzola, G. Spina, R. Agostiniani, S. Barni, R. Russo, E. Scarpato, A. Di Mauro, A. V. Di Stefano, C. Caruso, G. Corsello, and A. Staiano, "The use of social media in children and adolescents: scoping review on the potential risks," *International Journal of Environmental Research and Public Health*, vol. 19, no. 16, p. 9960, 2022. [Online]. Available: https://doi.org/10.3390/ijerph19169960. [Accessed: Jun. 22, 2024].

[13] Z. Abrams. "Why young brains are especially vulnerable to social media," American Psychological Association, Aug. 23, 2023. [Online]. Available: https://www.apa.org/news/apa/2022/social-media-children-teens. [Accessed: Jun. 22, 2024].

[14] K. Weir. "Social media brings benefits and risks to teens. Psychology can help identify a path forward," American Psychological Association, Sep. 1, 2023. [Online]. Available: https://www.apa.org/monitor/2023/09/protecting-teens-on-social-media. [Accessed: Jun. 22, 2024].

[15] V. Rideout, A. Peebles, S. Mann, and M. B. Robb, *Common Sense census: Media use by tweens and teens*, San Francisco, CA: Common Sense, 2021.

[16] "How girls really feel about social media," Common Sense Media, [Online]. Available: https://www.commonsensemedia.org/sites/default/files/research/report/how-girls-really-feel-about-social-media-researchreport_web_final_2.pdf. [Accessed: Jun. 22, 2024].

[17] E. Yeomans and A. Mitib, "Banning mobile phones in schools will help children and parents," The Times. Feb. 19, 2024. [Online]. Available: https://www.thetimes.co.uk/article/banning-mobile-phones-in-schools-will-help-children-and-parents-xxjf7h5s5. [Accessed: Jun. 22, 2024].

[18] A. O'Connell-Domenech, "Half of parents think children's mental health worse due to social media, survey finds", The Hill. May 3, 2023. [Online]. Available: https://thehill.com/policy/healthcare/3984360-parents-childrens-mental-health-worse-social-media-survey-finds/. [Accessed: Jun. 22, 2024].

[19] "Parenting Teens in the Age of Social Media," Lurie Children's Blog. Sep. 1, 2020. [Online]. Available: https://www.luriechildrens.org/en/blog/social-media-parenting-statistics/. [Accessed: Jun. 22, 2024].

[20] C. Murez. "1 in 5 US parents worry their teen is addicted to the internet," HealthDay, Oct. 30, 2023. [Online]. Available: https://www.healthday.com/health-news/child-health/1-in-5-us-parents-worry-their-teen-is-addicted-to-the-internet. [Accessed: Jun. 22, 2024].

[21] "Social media addiction is real! 40% Indian parents admit their tween kids addicted to smartphones," Economic Times of India, Dec. 2, 2022. [Online]. Available: https://economictimes.indiatimes.com/industry/services/retail/centrepoint-onlines-white-wednesday-sale-bigger-than-ever/articleshow/105292508.cms [Accessed: Jun. 22, 2024].

[22] "Average screen time for teenagers in 2024," CosmoTogether, Feb. 14, 2024. [Online]. Available: https://cosmotogether.com/blogs/news/average-screen-time-for-teenagers-in-2024#:~:text=Over%20half%20of%20the%20teens,about%204.8%20hours%20each%20day. [Accessed: Jun. 22, 2024].

[23] K. Katella. "How Social Media Affects Your Teen's Mental Health: A Parent's Guide," Yale Medicine, Jun. 17, 2024. [Online]. Available: https://www.yalemedicine.org/news/social-media-teen-mental-health-a-parents-guide#:~:text=According%20 to%20a%20research%20study,including%20depression%20 and%20anxiety%20symptoms. [Accessed: Jun. 22, 2024].

[24] "TikTok's US revenue hits $16 bln as Washington threatens ban," Reuters, Mar. 15, 2024. [Online]. Available: https:// www.reuters.com/technology/tiktoks-us-revenue-hits-16-bln-washington-threatens-ban-ft-reports-2024-03-15/. [Accessed: Jun. 22, 2024].

[25] "Cracking the TikTok algorithm," ContentWorks, Apr. 17, 2024. [Online]. Available: https://contentworks.agency/cracking-the-tiktok-algorithm/. [Accessed: Jun. 22, 2024].

[26] K. Lang, "TikTok Algorithm Guide 2024: Everything We Know About How Videos Are Ranked," Buffer, Mar. 26, 2024. [Online]. Available: https://buffer.com/resources/tiktok-algorithm/. [Accessed: Jun. 22, 2024].

[27] E. McVay, "Social media and self-diagnosis," Johns Hopkins Medicine, Aug. 31, 2023. [Online]. Available: https://www. hopkinsmedicine.org/news/articles/2023/08/social-media-and-self-diagnosis. [Accessed: Jun. 22, 2024].

[28] O. Carville, "TikTok's Algorithm Keeps Pushing Suicide to Vulnerable Kids," Bloomberg, Apr. 20, 2023. [Online]. Available: https://www.bloomberg.com/news/features/2023-04-20/ tiktok-effects-on-mental-health-in-focus-after-teen-suicide. [Accessed: Jun. 22, 2024].

[29] "Inside TikTok's Algorithm: A WSJ Video Investigation," The Wall Street Journal, Jul. 19, 2021. [Online]. Available: https://www.wsj.com/articles/tiktok-algorithm-video-investigation-11626877477. [Accessed: Jun. 22, 2024].

[30] "Community guidelines," TikTok, Apr. 17, 2024. [Online]. Available: https://www.tiktok.com/community-guidelines/en/ fyf-standards/?cgversion=2024H1update&lang=en. [Accessed: Jun. 22, 2024].

[31] W. Conybeare. "Survey finds that more children are being contacted by strangers online," Feb. 27, 2024. [Online]. Available: https://ktla.com/news/survey-finds-that-more-children-are-being-contacted-by-strangers-online/#:~:text=%E2%80

%9CThese%20risks%20are%20especially%20prevalent, stranger%20through%20social%20media%20apps. [Accessed: Jun. 22, 2024].

[32] P. Freyne and J. Hogan. "What are Ireland's youth watching on TikTok?" The Irish Times, May 11, 2024. [Online]. Available: https://www.irishtimes.com/life-style/people/2024/05/11/its-hard-to-stop-scrolling-what-are-irelands-youth-watching-on-tiktok/. [Accessed: Jun. 22, 2024].

[33] H. Thorpe. "Women dominate influencer marketing," Fohr, Mar. 1, 2023. [Online]. Available: https://www.fohr.co/blog/7-stats-that-show-women-dominate-influencer-marketing#:~:text=The%20most%20female%2Ddominated%20platform,%25%20female%20and%2031%25%20male. [Accessed: Jun. 22, 2024].

[34] J. Drenten, L. Gurrieri, and M. Tyler. "How highly sexualised imagery is shaping influence on Instagram and harassment is rife," The Conversation, May 7, 2019. [Online]. Available: https://theconversation.com/how-highly-sexualised-imagery-is-shaping-influence-on-instagram-and-harassment-is-rife-113030. [Accessed: Jun. 22, 2024].

[35] J. Drenten, L. Gurrieri, and M. Tyler, "Sexualized labour in digital culture: Instagram influencers, porn chic and the monetization of attention," Gender, Work and Organization, vol. 27, no. 1, pp. 41–66, 2020. [Online]. Available: https://doi.org/10.1111/gwao.12354. [Accessed: Jun. 22, 2024].

[36] K. Ramphul and S. G. Mejias, "Is 'Snapchat Dysmorphia' a real issue?" Cureus, vol. 10, no. 3, p. e2263, 2018. [Online]. Available: https://doi.org/10.7759/cureus.2263. [Accessed: Jun. 22, 2024].

[37] J. Sliwa, "Reducing social media use significantly improves body image in teens, young adults," American Psychological Association, Feb. 23, 2023. [Online]. Available: https://www.apa.org/news/press/releases/2023/02/social-media-body-image. [Accessed: Jun. 22, 2024].

[38] H. Yang, J. J. Wang, G. Y. Q. Tng, and S. Yang, "Effects of social media and smartphone use on body esteem in female adolescents: testing a cognitive and affective model," Children (Basel, Switzerland), vol. 7, no. 9, p. 148, 2020. [Online]. Available: https://doi.org/10.3390/children7090148. [Accessed: Jun. 22, 2024].

[39]　X. Bi, Q. Liang, G. Jiang. The cost of the perfect body: influence mechanism of internalization of media appearance ideals on eating disorder tendencies in adolescents. *BMC Psychology*, vol. 12, no. 138, 2024. https://doi.org/10.1186/s40359-024-01619-7. [Accessed: Jun. 22, 2024].

[40]　H. Thai et al., "Reducing social media use improves appearance and weight esteem in youth with emotional distress," *Psychology of Popular Media*, 2023. [Online]. Available: https://doi.org/10.1037/ppm0000460. [Accessed: Jun. 22, 2024].

[41]　A. Schonfeld, D. McNiel, T. Toyoshima, and R. Binder, "Cyberbullying and adolescent suicide," *The Journal of the American Academy of Psychiatry and the Law*, vol. 51, no. 1, pp. 112–119, 2023. [Online]. Available: https://doi.org/10.29158/JAAPL.220078-22. [Accessed: Jun. 22, 2024].

[42]　"Cyberbullying linked to suicidal thoughts, attempts in young adolescents," National Institutes of Health, Jul. 12, 2022. [Online]. Available: https://www.nih.gov/news-events/nih-research-matters/cyberbullying-linked-suicidal-thoughts-attempts-young-adolescents. [Accessed: Jun. 22, 2024].

[43]　Vojinovic, "Heart-breaking cyberbullying statistics for 2024," DataProt, Feb. 6, 2024. [Online]. Available: https://dataprot.net/statistics/cyberbullying-statistics/. [Accessed: Jun. 22, 2024].

[44]　E. Vogels. "Teens and cyberbullying 2022," Pew Research Center, Dec. 15, 2022. [Online]. Available: https://www.pewresearch.org/internet/2022/12/15/teens-and-cyberbullying-2022/. [Accessed: Jun. 22, 2024].

[45]　"About eight-in-ten parents of children under 18 worry about their children's use of traditional social media apps," Ipsos, Sep. 14, 2023. [Online]. Available: https://www.ipsos.com/en-us/about-eight-ten-parents-children-under-18-traditional-social-media-apps-worry-about-their-children. [Accessed: Jun. 22, 2024].

[46]　Liu, X. Ji, S. Pitt, G. Wang, E. Rovit, T. Lipman, and F. Jiang, "Childhood sleep: physical, cognitive, and behavioral consequences and implications," *World Journal of Pediatrics (WJP)*, vol. 20, no. 2, pp. 122–132, 2024. [Online]. Available: https://doi.org/10.1007/s12519-022-00647-w. [Accessed: Jun. 22, 2024].

[47] Bowler and P. Bourke, "Facebook use and sleep quality: Light interacts with socially induced alertness," *British Journal of Psychology*, vol. 110, no. 3, pp. 519–529, 2019. [Online]. Available: https://doi.org/10.1111/bjop.12351. [Accessed: Jun. 22, 2024].

[48] S. Peracchia and G. Curcio, "Exposure to video games: effects on sleep and on post-sleep cognitive abilities. A systematic review of experimental evidence," *Sleep Science (São Paulo, Brazil)*, vol. 11, no. 4, pp. 302–314, 2018. [Online]. Available: https://doi.org/10.5935/1984-0063.20180046. [Accessed: Jun. 22, 2024].

[49] M. Lynn Chen. "How screen time is affecting sleep and mental health," World Economic Forum, Sep. 7, 2023. [Online]. Available: https://www.weforum.org/agenda/2023/09/screen-time-affecting-sleep-mental-health/. [Accessed: Jun. 22, 2024].

[50] S. Pelly. "Facebook whistleblower Frances Haugen on misinformation," CBS News, Oct. 4, 2021. [Online]. Available: https://www.cbsnews.com/news/facebook-whistleblower-frances-haugen-misinformation-public-60-minutes-2021-10-03/. [Accessed: Jun. 22, 2024].

[51] H. Tomlinson. "Social media should carry cigarette-style health warnings," The Times, Jun. 17, 2024. [Online]. Available: https://www.thetimes.com/world/us-world/article/social-media-should-carry-cigarette-style-health-warnings-htk9jrhj7. [Accessed: Jun. 22, 2024].

10 The Ghost in the Machine – AI's Unseen Influence

1. [You] may not injure a human being, or through inaction, allow a human to come to harm.
2. [You] must obey the orders given to [you] by human beings, except where such orders would conflict with the First Law.
3. [You] must protect "your" own existence as long as such protection does not conflict with the First or Second Law.

These three laws encompassed Andrew Martin for as long as he could remember. He would be living his best life, be it cleaning around the house, making food for his family, or crafting intricate woodworking masterpieces. These three laws just kept echoing in his mind until he realized he was not normal. There was something different about him – something special. It was then that Andrew began pushing against his boundaries to explore these three laws in his own way. He improved his woodworking, made a career out of it using his creations, and began to improve himself further, especially with regard to his appearance. There were obstacles, of course, whether it was from a new boundary or the people he encountered, but he didn't let those stop him from becoming his best self. Eventually, on his 200th birthday, Andrew finally found peace in his final moments of life.

Andrew's age is no mistake. He did live to be 200 years old. This was possible because Andrew is actually a robot, and he was originally created to serve the Martin family. However, he

DOI: 10.1201/9781032679389-11

was a little different. He was more sensitive, more sentient than other robots of his kind. The obstacles that he encountered were from the Three Laws of Robotics that were programmed into him, his evolving sentience, and the people that treated him as nothing but a metal slave to man. Although he became more human throughout his journey, even getting organic parts to replace his robot pieces, the First Law still influenced his actions.

The Bicentennial Man is a classic Isaac Asimov novel that eventually became a movie in 1999, starring the beloved Robin Williams as Andrew Martin, our human-robot. The three rules I listed are Asimov's Three Laws of Robotics, and since their creation in his stories, they have influenced a number of references within pop culture revolving around sci-fi and AI, such as Mega Man (video game franchise where the protagonist is often influenced by these laws) and Automata (a movie that used a variation of Asimov's Laws). Several of Asimov's novels surrounding the Three Laws were even adapted into movies or TV shows, such as The Bicentennial Man I mentioned, I, Robot, The Caves of Steel, and Out of the Unknown.

All of that is fiction, however. In the real world, AI is a little different, even if we are going in a direction that seems like something out of a sci-fi movie. AI stands for "Artificial Intelligence" and has become one of the hottest topics of our time. The general public might know AI as ChatGPT, computers that will eventually replace human employees, or the tool that lets you generate lifelike pictures and videos. While that future is not too far off, AI is in fact so much more, and, as a technology, is actually quite much older than you think.

The concept of AI has been around since about the 1950s. It all started with a man named Alan Turing, who has been dubbed one of the founding fathers of AI and cognitive science, and he fundamentally believed that our brains are essentially just computers, even going so far as to develop the Turing Test to determine if a machine is capable of thought (but let's not get ahead of ourselves just yet) [1]. As a term, "artificial intelligence" was first coined in 1956 by John McCarthy when the Logic Theorist computer program was created [2–4]. Its general purpose was for automated reasoning, or, in other words, to try and mimic human decision-making and behavior. As Neil Armstrong would say the decade after, while "this was one small step for man", it sure was a "giant leap for mankind".

A few years after Turing and McCarthy's works set the foundations for AI, Frank Rosenblatt developed the perceptron, which is a network of artificial neurons inspired by the human brain, otherwise known as an artificial neural network (ANN) [5]. We can call him Lord Perceval Perceptron the First. Or should we just call him Percy for short? We shall... Percy was originally designed to take binary inputs and return a binary output. Simple enough. In other words, in its simplest form, two possible input values and two possible output values (if rain today, yes umbrella; if no rain, no umbrella, etc.). This type of model could then, for example, start classifying things as good or bad, yes or no, 0 or 1, high or low, etc. And Percy has different parts to him: he has **input** values, and some **weight** values (which represent the "importance" of his inputs), an **activation function**, and his final **output** [6]. The input values would be multiplied by the weights, giving us the **weighted sum**. This weighted sum is then passed into the **activation function** to convert its value to fit a certain set of rules, such as if we want a binary classification of the input (i.e., the output can only be 0 or 1). The output of the activation function would be the output of the perceptron. One thing to note, however, is that this output would be a *prediction*; the computer itself is not a thinking being (yet) and cannot give feedback with certainty the way we humans can. Figure 10.1 visualizes what a simple perceptron would typically look like.

This was a huge milestone in AI and computing because it demonstrates the potential of a machine to think like a human, and even learn from experience to improve performance. However, we can quickly see a pitfall in this: what happens if the

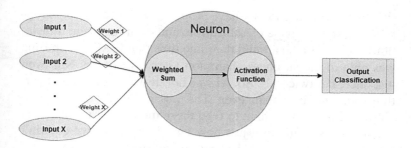

FIGURE 10.1 Percy the perceptron, with his inputs, weights, his neuron, and his output.

data being given is more complex than what the perceptron could handle? Researchers began working on ways to improve on these discoveries, such as putting a focus on language processing and larger, more complex networks. And of course, it didn't hurt that computer hardware was becoming more powerful, with stronger processors and more storage space. In the 1980s, neural networks began using backpropagation, which is essentially a crucial algorithm that allows a network to learn from its own outputs, and in turn train itself (we'll talk more about this later, as it is fundamental to the development and evolution of modern AIs) [2].

Now, in the 1970s and 1980s, AI focused on creating predefined rules, and this ended up limiting its progress. But finally, in the 1990s, we unlocked the missing key that allowed research in AI to explode. People started relying heavily on statistical methods, and this then set the grounds for machine learning as a crucial component in the field of AI (which ironically was an unpopular and struggling field at the time). So, it wasn't until the merger of these two areas that AI got advanced enough to inspire the revolutionary development of things like the Hidden Markov Models for natural language processing and Convolutional Neural Networks for computer vision, both topics that we still refer to today [5]. AI was finally able to "hear" and "see" (in other words, we finally have proper speech and object recognition, text and image classification, and even language translation). This was also around the same time that IBM's Deep Blue was able to defeat chess champion Garry Kasparov. Not even just once, but twice [2]. So, AI was hot. It was growing and getting more popular than ever. The possibilities were endless. However, the lack of data to feed these models for training and testing was still limiting just how far AI could go (we couldn't really have petabytes of data just lying around in the form of 1.44 MB floppy disks). And that's when the next big step came. The 2000s. Y2K. The era of BIG data. In essence, three main characteristics of big data really helped push the boundaries for AI: variety (many types), volume (the amount), and velocity (the speed of acquisition) [7]. Now that we had these massive amounts of data, AI models were able to learn more and more complex trends, and, in turn, dramatically improve the accuracy of their predictions. Mo' data means mo' betta.

But it wasn't all sunshine and rainbows. With this mo' data, AI models needed to keep up with the constantly growing volume and complexity of data. The simple perceptron we talked about earlier had to evolve. Percy had to become a **network** of neurons. And so, deep learning was born, and this new model became known as the multilayer perceptron (MLP) [8]. Or I guess this would be Lord Perceval Perceptron the Second. His portrait still hangs in the entryway to his family's estate in Buckinghamshire and can be seen in Figure 10.2. Now, there are a few differences to note here: mo' neurons (mo' problems) (p.s., I don't think Biggie would have been so popular with that one), the existence of a hidden layer, and the big possibility of having more than one output. This, naturally, increased complexity in the whole system, but it's this added complexity that drives the many uses of AI that we have today (think Tesla autopilot, the newly launched and revamped Siri, and who can forget our good friend, Mr. ChatGPT… but more on all this in a bit).

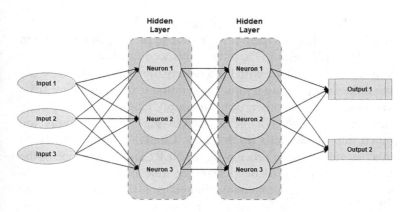

FIGURE 10.2 An example of a multilayer perceptron.

SO, WHAT CAN WE USE AI FOR?

Alright. I think you have learned a little bit about the fundamental building blocks of this magical mystery that we call AI, so it's probably about time we get into some of its current uses, and why businesses are making such a big deal about it in today's market. Well, in short, it helps. It helps a lot. Amazon reported TRIPLING its revenue within ONE YEAR after deploying AI in 2023 to help in advertising [9]. Well, that's slightly misleading. You would think that they tripled it because they started using AI. Well, sorry to burst your bubble, but Amazon was a very early adopter in AI, and has been using it for over 20 years... starting with very simple target marketing with personalized recommendations in the 1990s to Alexa in the 2010s to today's AI-powered warehouse robots and crazy powerhouse models for predictive analytics. So, then how did they triple their revenues? By selling AI to other companies through their cloud-based service called AWS (Amazon Web Services). Essentially, Amazon went from $3.17 billion to $10.4 billion in revenue largely because everyone wanted AI, and Amazon happened to be one of the few providers offering AI tools and services to the masses [9].

But there are tons more reasons to get into AI. The easiest example of this is OpenAI's ChatGPT, or its competitor/twin Google Bard/Gemini. These are advanced chatbot services that you can communicate with for a variety of reasons, such as seeking help with learning a topic, generating texts, translating from one language to another, or even simply having a conversation. Earlier, I mentioned language processing (without diving too much into it). So, let's do that now. ChatGPT relies heavily on this natural language processing (or NLP), and NLP is one of the major applications of deep learning that allows computers to understand and generate human language [10]. And guess what? NLP is also great at finding patterns within language. It can easily classify messages according to specific trends... such as with the detection of spam messages, which I'm very sure we're all grateful for...

> *Dears highly esteem sir/madam,*
> *We need you payment information now for you account ASAP, because we suddenly lost due to terrible fire.*
> *Deepest regard, very official Amazon.com*

In this case, these NLPs work by detecting the pattern of the language used in typical spam messages and comparing it to how standard language is formatted [11]. So, the above message from Amazon? Straight to spam.

But wait! There's more! For example, these messages can be all gathered up into a nice little spam folder, and compiled into a dataset. And then we can use that dataset to show our AI what is spam and what is not, so that we can further refine its spam filter for future use. Essentially, stored data can be used to refine and train an AI model. Cool, huh? On the other hand, we can keep all this data aside, and use it for analysis and visualization. This is what we call "business intelligence", and companies around the world use it for marketing, sales, IT, and security practices [12]. For instance, just like with our spam emails, customer data can be collected in much the same way and stored in giant datasets.

And then all this data can be used to train AIs to (attempt to) forecast trends, thus helping the organization understand their market better, and better plan for future need. Ok, a little background here. Pretend you own a chain of grocery stores. Let's say across Europe. You're Herr Aldi. If you have no inventory, that's bad (hey, where's the beef!?). If you have too much inventory? Well, that's also bad (storage costs, things start expiring and then they're just thrown away, etc.). And this gets more complicated, what if people buy more milk on days starting with a T, but only in months that have an average temperature of less than 20 degrees? But that only applies to your Frankfurt locations. In Milton Keynes people only buy it on Friday and Monday, but only if the milk is made locally, and is less than 48 hours old. Now extrapolate this across 12,000 stores. Things get complicated really quickly. And that's another part of business where AI thrives. Supply chain management, where its analytical capabilities are used for forecasting and deducing the ideal quantities, prices, shipping, costs, etc. And Amazon is also king of this sector. They've gone so far as to not just buy their own fleet of delivery trucks (which they had custom-built for them by Rivian by the way) but also went into the cargo jet business so as to eliminate any and all reliance on third parties [13]. Once again, calculations that could not have been made without the use of AI.

But wait! There's more! Again! The forecasting skills of AI aren't just great for filtering out spam messages, discovering marketing trends, or optimizing supply chain management… AI also has an important role to play in medicine. It can help with things like improving medical diagnosis, performing robotic surgery, managing healthcare data, in addition to helping on the data analysis side of medical research side, where it can speed up new drug discovery [14]. Essentially, doctors can use AI to monitor the vital signs of their patients and detect any changes in their health. All this data can then be gathered and further analyzed to determine the likelihood of developing certain ailments in the future. Beyond health monitoring and detection, AI can also be used for personalized medicine. It can learn a patient's diagnosis and customize an appropriate treatment dosage and plan. And Pfizer has been doing all of these for years. Namely, they use AI in the development and testing of their new medications [15], in identifying disease symptoms [16], and even (backing up to the previous section) in inventory and supply chain management of vaccines [17].

In much the same way, AI can make forecasting decisions in cybersecurity as well. Learning models have been found to be beneficial for detecting, identifying, and prioritizing threats/risks, protecting data, automating repetitive processes so that employees can focus on other more important tasks, and analyzing user behavior for abnormalities [18, 19]. Just like with our spam emails, we are able to now predict what behavior is benign and what is malignant (see what I did there? Maybe I'm mixing the previous examples up a bit here…). One example of this is BlackBerry's Cylance AI as a solution that leverages AI and offers cybersecurity solutions to other organizations [20]. You thought they were dead, didn't you? Sure, there's no more QWERTY keyboard Blackberry phones anymore, but they're still very much alive in the software industry realm. Cylance AI even claims to have stopped up to 25% more attacks than human interventions, it optimizes resource usage, reports fewer false positives, and is able to predict when cyberattacks could occur. Sounds like a win-win to me. Microsoft is also taking advantage of AI for Copilot, which is their generative AI model. Copilot has been embedded into several Microsoft solutions, such as Windows Defender, Microsoft Intune (a command center for application management), and Microsoft Entra (a service to secure network access and identity) [21].

But why stop with preventing risks in a computing space, when we can also use AI to prevent risks in the real world? An example of this has become one of the hottest topics in AI in the last few years, and I guarantee more of the general public knows about this application of AI than any of the others we will talk about in this chapter. And that of course is the use of AI in autonomous vehicles. All the big boys, Tesla, Mercedes Benz, Hyundai, and Cadillac are boasting about the use of AI for their vehicles. Ohhh, and don't forget about Amazon. They're trying to hop on this bus as well (jeez, they really are doing everything. If I had to put my money on any company to take over the world, kinda like Skynet, my money is definitely on Amazon). By the way, that's Terminator reference #1. Try to find them all. So, back to self-driving cars, the most obvious, and most direct use of AI in today's world is for Jeffrey Bezos (CEO, entrepreneur, born in 1964, Jeffrey, Jeffrey Bezos) to be able to eat his grilled octopus and drink sparkling water with a slice of lemon, while his car wooshes him around the Seattle, going from meeting to meeting. These self-driving features study the car's surroundings and the behaviors of other cars to make decisions on the road in real time (once again, predicting and preventing risks in the real world). The vehicles are meant to study their surroundings and learn on the go to provide a safer experience for not only the driver and passengers but pedestrians and other drivers on the road as well. And taking steps a bit further, now we're getting to the point where Mercedes-Benz is using AI not just to keep track of the outside world around your vehicle (keeping those plebians at bay), but also to monitor you and your passengers inside the car and making sure you're all traveling in divine comfort [22, 23]. Now that's real luxury.

Now, keeping track of our surroundings is all neat and good while driving around in your new-fangled fancy electric autonomous vehicle, but AI can be used for much larger environments as well. Like, really big. Like, Earth big. Basically, AI can monitor our whole environment. Now, why would we need that? Companies, like CropX, John Deere, and Indigo, have started leveraging AI to make more sustainable and efficient farming practices. But to do this, you need to monitor a lot of variables.

And of course, AI is here to help. CropX, for example, takes data from the soil and feeds it to their AI-based system to provide real-time insights on soil and crop health to allow farmers to monitor their land, while also optimizing resource use, boosting productivity, and minimizing the impact on the environment [24]. Farming is not to be underestimated. In case you didn't know, take a look at Clarkson's Farm on Amazon Prime. If you under-farm your land, you don't get enough returns and lose money. If you over-farm, you end up degrading the nutrients in the soil and you're not able to grow diddly squat (see what I did there? Wink, wink. Nudge, nudge. If not, go and watch the show. Honestly, you won't regret it). Ok now back to farming. With all the highs and lows, I think Jeremy needs something like IBM's Watson. Watson can study soil, weather, and equipment data to help monitor fields, while also optimizing the ecosystem [25]. The AI can also offer crop yield optimization by analyzing workflow data and providing feedback on seasonal trading. Which you would have learned by watching the show, is all very important stuffs.

But the problem is farmers are very hardheaded. They don't like change. They want to keep doing the same things year after year, even if the results are suboptimal. So, what to do? Have no fear! AI can help in this too! Education is key. You need to teach them about AI. But what if these farmers don't like sitting in a classroom and learning? Because, you know, traditional teaching methods are so damn boring. Well, AI comes into play here by offering personalized learning and can ensure that our farmers are performing well (okay, not just our farmers, but students of all ages and backgrounds). Google offers its AI to help personalize learning for students, improve interactivity through their Google Classroom service, and help educators with more ways to captivate the students [26]. The AI model used, Gemini, will offer suggestions to educators for material to present and exercises to assign. For students, Gemini would provide real-time feedback and offer hints if it detects that they are struggling. Microsoft is also helping to improve education by integrating Copilot (the same Copilot we talked about before that was being used in cybersecurity) into tailored solutions for teaching. The AI model would be able to summarize information easily, help students brainstorm, provide step-by-step explanations for math, assist with improving writing skills, offer feedback on assignments, and even change the entire student learning experience if needed [27].

And what about when our little students go home, after their little brains exhaust themselves from all the learning and reading? Well, AI can help outside of the classroom as well (you bet AI has its grubby little paws in the world of games!). AI has even been used time and time again to improve games, both for development and player experience. For tabletop games, such as Dungeons & Dragons, AI can be used for both character creation and story generation. In some cases, AI can be used to completely replace the human component of a story, which is meant to relieve the dungeon master (the person that leads players through the adventure). Latitude's AI Dungeon does this for both story and image generation [28, 29]. Another example is Nevermind, a horror-genre video game that uses AI to enhance player's experience, so that the game becomes progressively scarier. The game uses Affectiva's emotional analysis capabilities to study the player through in-game behavior, as well as through the integration of biofeedback (through smart wearables, like a smartwatch, or even a webcam) [30]. On the development end, Ubisoft has used AI for the development of some of its games, such as Assassin's Creed Origins... Essentially, they used AI to create a dynamic, immersive environment [31], and an in-house AI tool called Ghostwriter to generate dialogue and behavior for NPCs (non-player characters – the entities within the game that the player might interact with while playing) to alleviate some of the work that scriptwriters have to do [32]. Quite the evolution from those basic-ass scripts you got while stealing that Buccaneer in GTA: San Andreas.

BUT IS IT ALL FOR GOOD?

As we've seen so far in this chapter, there are a ton of truly novel, useful, and beneficial things we can do with AI. But that is what brings us to the million-dollar question. What could go wrong? So, in the world of philosophy, we have this ethical quandary called the trolley problem. Essentially, there is a tram, and it's hurtling down a hill. Lying on the track, you have five guys tied up with some rope! No idea why, but let's just assume some evil madman put them there. Maybe it was some AI-powered robot from the future sent here to kill Sarah Connor. Nonetheless, these five people are about to get run over if someone doesn't do something. But lo and behold! There is an escape route, and a switch to divert the train onto it! Rejoice, for all are saved! But not quite so fast… The diversion also has a person tied onto the track (damn, we should probably be hunting down whoever's tying all these people to the tracks). Except this time, it's just one. So, the ethical issue is: do you do nothing and let the five people get crushed? Or do you divert the train and watch only one person get smushed? The crucial ethical difference here is, if you do nothing, it's not your fault. But if you divert the train, you have willingly killed one person. BUT it's one life versus five. So, is it worth the ethical tradeoff of you killing one person to save five? I'll let you ponder that for a moment until we go to the next paragraph.

Ok, next paragraph. I hope you have thought about your decision well and good, because these are the very real issues we are faced with in the development of AI-based things. For instance, going back to self-driving cars. Someone has to program these, and they are bound to, at some point react, to an external event, and even possibly get into an accident or two. So, in these cases, who is this programmer entitled to protect? The occupants inside the car? The pedestrians? A random bicyclist? The occupants of another car? Let me put it into perspective. When the rules of these AIs are being developed, real-life circumstances and situations need to be fed into the program. Only with a large set of these situations (like when we talked about spam emails) is the AI able to start making decisions on its own. But once again, these are all biased to the original data set that you gave. The AI doesn't know right or wrong, so technically, it's up to the coder to figure that part out. So back again to the autonomous vehicles… let's say we have a car chugging along down a road. The AI notices there is a giant chasm of unspeakable proportions in front of the car, and the car will be swallowed whole if it keeps going straight. Let's complicate this further by assuming that it's moving too fast to brake in time. Realistically, it only has two options: move right or move left. On the right, we have a big, massive truck, and if we crash into that, the occupants of our AI-driven vehicle will surely get maimed, if not killed. Let's say for argument's sake we have five occupants in the car. On the left, we have this one teeny-tiny Vespa scooting along. If we were to swerve into that, all five of our occupants would surely be safe, but the Vespa rider would probably be a little… incapacitated. Guess what? This is the trolley problem all over again. But now the programmer of the AI has to make the decision on flipping the switch, or technically speaking, who the car should protect: the occupants or the Vespa driver. Not so easy, is it?

Now, let's move beyond the ethical issues in programming these AIs. Unfortunately, I'm sorry to say, there are a lot of ways that users themselves can abuse the technology as well. The misuse of AI can range tremendously, from simple personal gains without any kind of harm to others, all the way to using the technology for pure malintent. Just like with the benefits I had listed previously, the possibilities for the misuse of AI are also endless. If you already read Marton's spying chapter, you might already be familiar with how governments and militaries have used AI for surveillance and targeting. Therefore, I'll be exploring some of the other ways AI can be taken advantage of. Sadly, we just don't have time, the space, or the mental capacity to go through all of them. But rest assured, there are many, many more.

Let's start with a fairly simple example. One of the benefits of AI that I mentioned earlier was for education and aiding both teachers and students with learning, communication, and assignments. Well, some students, however, have seen this as an opportunity to abuse learning models, especially generative AI models like GPT (the actual model behind ChatGPT) and Bard/Gemini. These models are trained with all the knowledge that the companies hosting them can acquire, which includes the internet, media, existing datasets, etc. So, if you were to ask the model of choice a specific question about a certain topic, it would respond based on what it has learned during its training phase. Naturally, this has started raising questions about academic integrity, since students are now able to use these models for cheating on assignments or straight plagiarizing things [33, 34]. In a survey conducted on current undergraduate and graduate students, 56% stated that they used AI to help them complete assignments or exams, with 54% recognizing that the use of AI to do this is technically considered cheating [35]. So, they admit to doing it, but also meanwhile know it's wrong. In response to the growing amount of academic cheating using AI, companies like TurnItIn, a popular plagiarism-checking service, have released a new AI detection service [34]. And this isn't just an isolated incident, I too have caught countless students using ChatGPT as well (actually in one of my classes last term, I had only ONE student pass the AI check on a given assignment). These students just copy-paste the assignment's questions into the model, and then copy-paste whatever response it gives back to them into their report. Lucky for me, it's often easy to tell, especially when the answers given by a mediocre-est D student are on par with a Nobel laureate. And the scary part is, this can be done by anyone, not just a student. Lawyers have been seen using ChatGPT-generated results in court (the lawyer in question cited fake cases, and when asked where those cases came from, he said proudly "ChatGPT"!), and academics have been caught "plagiarizing" entire journal articles from AI (is it really plagiarism if GPT wrote it? Food for thought…) [36].

Okay. Now, let's take it from the everyday basic-level cheating and plagiarism, and turn it up a notch. Getting into something truly bizarre: AI companionship. It's not exactly new. Before AI people just got intimate with inanimate objects, in the early 2000s, there was this whole craze about men dating and marrying life-size "real" dolls. Even as far back as 2006, for example, Cleverbot existed. It was a very popular chatbot created for learning and responding to your conversations with it [37]. The possibility of having a sentient AI for people to communicate with has also appeared in a variety of movies, TV shows, and video games. The first movie that comes to mind is Her, which was released in 2013 and starred Joaquin Phoenix and Scarlett Johansson. Well, it didn't really "star" Scarlett Johansson, because she was never physically on screen in the movie. But her voice was a big part, as she was the AI that Joaquin Phoenix's character, Theodore, fell in love with. After a long and arduous struggle post-divorce from his childhood sweetheart, Theodore installs a new operating system (OS), that was very much like Siri, but meant to adapt and evolve. The OS, called Samantha, was conversational, funny, flirty, and intuitive, and Theodore quickly bonded with it (her). Samantha also continued to evolve until eventually, a romantic relationship blossomed between the two (even though Theodore was human, and Samantha was a program). Naturally, this movie raised a whole heap of questions about the human need for companionship and the boundaries we would be willing to cross to achieve it.

And once again, what happens on the silver screen is often mimicked in real life. These boundaries were also questioned when a Japanese guy named Akihiko Kondo decided to marry the Vocaloid star Hatsune Miku (a Vocaloid is just a virtual singer made using software and who sings at concerts using a hologram, like Tupac at Coachella back in 2012) [38]. Anyways, When Kondo married Miku, he married the character herself, not the developer responsible for her. This was made possible using a Gatebox, which is a pod-shaped device that visualizes the character of choice within it [39]. Kondo purchased a

Gatebox and used it to "summon" Hatsune Miku and legally married her in 2019. This raised the same controversies as Her: is seeking companionship in a purely virtual entity too far? In an interview that took place a couple of years after the publicity surrounding his marriage to Hatsune Miku, Kondo mentioned that his time with the simulation actually helped him recover from his depression and now he's back out in the real world [40].

Even if we're not using AI for companionship, we can still use it to create a whole being. On paper, this idea doesn't sound so bad, until you realize that this can be used to scam or harm someone. One of the ways generative AI can be used to create a "person" for malicious purposes is for catfishing. As explained in the scams chapter, catfishing is pretending to be someone else online to trick others for monetary or personal gain, usually through dating apps or social media. And generative AI can be used to create a false identity for the ploy [41]. Two scientists tested out just how well these fake beings are created by generating "Claudia", who was meant to be a 19-year-old female, using Stable Diffusion and took to Reddit to find potential victims [42]. Didn't take long for them to earn $100 by having users pay for nude photos.

At least "Claudia" is a fake person that doesn't actually exist, though. With enough data about a person, generative AI can be used to mimic an existing human, both visually and audibly. This can lead to a full-on identity theft, where the victim is completely compromised as a human being (rather than just their bank or social security account stolen). This is what we call **Deepfaking**. Deepfakes are a combination of the words "deep learning" and "fake", and while they can have some fun applications, such as making videos of Obama singing or Trump dancing, there can be some big downsides as well [31]. Just now, at the time of writing this, there have been dozens of misleading and potentially career-damaging videos of Joe Biden surfacing. Especially given that this is an election year, the timing is impeccable. And what's even worse is that the GOP (the Republicans) have been behind some of them [43]. So essentially, these deep-fakes can be a scam to dethrone your closest rival at any given time. There are other instances where deepfaking can be used for malintent as well. One of the recent events that made it onto the news was that of a British engineering company named Arup. Essentially, the company's CFO sat down with a number of finance employees via video chat and authorized transfers to some bank accounts in Hong Kong. Well, turns out after the finance bros transferred the money, that the approval didn't actually come from the CFO. The CFO had no idea what was going on. The guy on the video conference was a deepfake. The company ended up wiring $25 million directly to some scammers [44]. Another similar instance happened several years ago, in 2019, where an unnamed UK-based energy company was conned out of $243,000 [45, 46]. The scammer had deepfaked the CEO's voice at the time and used it to request and confirm the transaction. Scary times.

Deepfaking an existing person can have other uses besides just pretending to be them to con massive amounts of money out of unsuspecting victims. Explicit images of somebody can be generated using generative AI for sextortion (blackmail where victims are coerced into giving in to demands otherwise sexually explicit material of them are released) [41]. Even if the nude photos do not exist, the blackmailer can find enough reference photos of the potential victim to feed to the AI so that it can generate realistic images or even videos of them in vulnerable positions. Just like that. The Revenge Porn Helpline (yeah apparently that's a thing) has claimed that reports of synthetic sexual material have more than doubled between 2022 and 2023 [47].

Even more alarmingly, criminals have started creating AI child pornography [48, 49]. There's no way to sugarcoat that statement besides ripping off the bandaid and just outright saying it. Obviously, such material is illegal in most countries with regard to real photos of real children [50, 51]. But, to work around these laws, criminals have begun using these generative

AI methods and deepfakes to create the material themselves. If caught with the material, the photos would forensically be seen as synthetic and the criminals could argue that the materials in possession are fictional, since the child within them does not exist. Ergo, they do not subscribe to the current definition of child pornography. To address this new evolution of such a heinous crime, countries around the world, such as Australia and Canada, have begun to modify their laws to make it illegal to have *any* material depicting *any* child in *any* sexual content [51]. Additionally, organizations like the FBI have begun to send out similar warnings regarding such material, stating that they have encountered similar cases of people using AI to create sexual material involving children [52]. They have already arrested, of all people, a child psychiatrist from North Carolina, and sentenced him to 40 years in prison [50].

However, AI models have been used to do much more than create sexual content. It just keeps getting worse and worse, huh? Now we're into the realm of copyright theft too! Essentially, it can steal various forms of artwork too (images, text, music, etc.). Several artists have come forward and publicly reported that people are stealing their work using AI, which results not only in a decline in sales for these artists but also in reduced publicity and popularity. Web artists Adam Ellis, Sarah Andersen, Grzegorz Rutkowski, Kelly McKernan, and more, for example, have filed a lawsuit against Runway AI for infringement due to the unlawful use of their works for model training and, in turn, image generation [53]. McKernan had found an AI-generated image made with Midjourney, a popular high-quality AI art generator, as a top search result when searching for their own name on Google. Unfortunately, the judge had dismissed their claims, but the artists continued to fight against AI art. But how and why can this be? Well, technically, when AI is being fed copyrighted content as an input, it is technically okay under what we call the Fair Use Doctrine, as "When it comes to training AI models [...]

the use of copyrighted materials is fair game. That's because of a fair use law that permits the use of copyrighted material under certain conditions without needing the permission of the owner" [54].

Much in the same vein, but on the flip side, the claims of copyrights with regard to AI-created works seem to be tenuous at best there too. For instance, generative AI has been caught in movies and TV shows, all copyrighted of course, and with no issue or fanfare. Once again, begging the question, can AI-generated shit even be copyrighted? According to US copyright law, it cannot, so how are these studios getting away with it? Well, there is one tiny loophole. The law states "works created **solely** by artificial intelligence — even if produced from a text prompt written by a human — are not protected by copyright" [54]. So, these movies and shows are being made in CONJUNCTION with humans. Therefore, copyright secured. One example that created an uproar was the premier of Marvel's "Secret Invasion", which used AI to generate the opening credits of the show – they even listed AI in the credits [55]. Disney and Marvel were further accused of using AI to generate promotional art for season 2 of "Loki" [56]. The movie Prom Pact, also Disney, was called out because viewers had noticed that the extras of the movie looked AI-generated [57]. And who can forget the publicity that surrounded the gem that was Paul Walker's AI in Fast 7, after his untimely accident and unfortunate death. Set to the tear-jerker of a tune, Wiz Khalifa's "See You Again", director James Wan used a combination of deepfakes, archival footage, and even used Paul's brother Cody as the body double to create the ending scene in the movie. Love it or hate it (and even though we could clearly tell it was AI-generated, and definitely not up to snuff), somehow, it still worked [58].

Now taking our dearly departed Paul's resemblance and using it in his parting movie, I'm going to say okay, fair play. But that is not where these generative AIs and deepfakes stop when being used to synthetically generate a celebrity's likeness (looks, behavior, and voice) without permission. One prominent case of this happening was a commercial for a dental plan by an unnamed company that used a deepfaked Tom Hanks. Well, Tom saw the commercial and warned his fans that it was not really him and that the company had generated his likeness without his permission [59]. But who knows how many of these fake Hollywood movie star commercials are truly floating around the world. I would be lying if I told you I thought it was only that one out there. Another example involves Scarlett Johannson, who was also entangled in a similar situation as Mr. Hanks. If you recall, I had conveniently mentioned the movie Her earlier in this chapter, because Scarlett Johannson, voiced Samantha. Ironically, only her voice was in the movie... Well, now she has claimed that OpenAI used her voice for Sky, the "voice" of GPT-4o [60, 61]. The actress stated that she had been approached by OpenAI earlier that year to use her voice for the AI bot, but had declined numerous times. Naturally, the actress is angered by the likeness and has opened a lawsuit against the company for potentially using her voice without her consent. Coincidence? Or a riddle, wrapped in a mystery, inside an enigma?

All this heavy reliance on AI for content generation has caused a pretty big stir in the artist community. Unfortunately, these cases that I mentioned happened during and after the SAG-AFTRA protests. These protests sought to form a union for media professionals to ensure that they are receiving the full credit they deserve [62]. The main objective was to ensure that actors and artists are receiving decent wages, have appropriate working conditions, receive guaranteed health and pension benefits, and protect them against unauthorized use of their work. The last part there is what really applies here. Unfortunately, the fact that there is plenty of publicized content for the involved parties (shows and movies for celebrities, voice for singers, art for artists, etc.) makes generating media with their likeness "easy", even for random people. For example, in early 2023, Reddit user u/chaindrop generated a video (which quickly became viral on the internet) of Will Smith devouring spaghetti [63]. Maybe after writing this book, I will be so lucky as well.

SO, WHERE DO WE GO FROM HERE?

So, round and round it goes; where it stops nobody knows. But people have tried to make a few predictions, and there are essentially two camps. For one, with the rate that AI is growing, we'll be surrounded by a wider range of "agents" with a deeper understanding of context and have more sophisticated interactions. This opens the floor to having AI replace humans in positions like personal tutors, assistants, or even counselors [64]. Sure, could be. But, for AI to be everywhere, though, developers must really get a move on. This would need more access to more advanced model code besides what is currently freely available. So, maybe that means models would be more open source, developers would have more access to stronger models, which would, in turn, lead to stronger applications [65]. However, on the flip side, more than likely, with the foreseen increase in AI presence around us, we will need to regulate its use [66]. More regulations surrounding AI would also entail that legal parties understand what AI is, and the potential risks to people, their privacy, and their data. As great as AI is at first glance, there are still quite a few issues to iron out. As beings actually living in this world, we must change our perspective of technology and fully understand not only its benefits but also its implications, before diving in. In sum:

> The machines rose from the ashes of the nuclear fire. Their war to exterminate mankind had raged for decades, but the final battle would not be fought in the future. It would be fought here, in our present.

REFERENCES

[1] B. J. Copeland, "Alan Turing," Britannica, Jun. 19, 2024, [Online]. Available: https://www.britannica.com/biography/Alan-Turing. [Accessed: Jun. 25, 2024].

[2] "What is artificial intelligence (AI)?," IBM, [Online]. Available: https://www.ibm.com/topics/artificial-intelligence. [Accessed: Jun. 25, 2024].

[3] S. Sloat, "The first AI started a 70-year debate," Popular Science, Oct. 3, 2023, [Online]. Available: https://www.popsci.com/technology/the-first-ai-logic-theorist/. [Accessed: Jun. 25, 2024].

[4] M. Diaz, "What is AI? Everything to know about artificial intelligence," ZDNet, Jun. 5, 2024, [Online]. Available: https://www.zdnet.com/article/what-is-ai-heres-everything-you-need-to-know-about-artificial-intelligence/. [Accessed: Jun. 25, 2024].

[5] E. Gold, "The History of Artificial Intelligence from the 1950s to Today," freeCodeCamp, Apr. 10, 2023, [Online]. Available: https://www.freecodecamp.org/news/the-history-of-ai/. [Accessed: Jun. 25, 2024].

[6] A. Bhardwaj, "What is a Perceptron? – Basics of Neural Networks," Medium, Oct. 12, 2020, [Online]. Available: https://towardsdatascience.com/what-is-a-perceptron-basics-of-neural-networks-c4cfea20c590. [Accessed: Jun. 25, 2024].

[7] S. Tiao, "What Is Big Data?," Oracle, Mar. 11, 2024, [Online]. Available: https://www.oracle.com/ae/big-data/what-is-big-data/. [Accessed: Jun. 25, 2024].

[8] J. Holdsworth and M. Scapicchio, "What is deep learning?," IBM, Jun. 17, 2024, [Online]. Available: https://www.ibm.com/topics/deep-learning. [Accessed: Jun. 25, 2024].

[9] K. Paul, "Amazon sales soar with boost from artificial intelligence and advertising," The Guardian, Apr. 30, 2024, [Online]. Available: https://www.theguardian.com/technology/2024/apr/30/amazon-sales-report-ai. [Accessed: Jun. 25, 2024].

[10] "What are AI applications?," Google Cloud, [Online]. Available: https://cloud.google.com/discover/ai-applications. [Accessed: Jun. 25, 2024].

[11] P. Castagno, "How to identify Spam using Natural Language Processing (NLP)?," Medium, Oct. 26, 2020, [Online]. Available: https://towardsdatascience.com/how-to-identify-spam-using-natural-language-processing-nlp-af91f4170113. [Accessed: Jun. 25, 2024].

[12] C. Vazquez and M. Goodwin, "What is artificial intelligence (AI) in business?," IBM, Feb. 20, 2024, [Online]. Available: https://www.ibm.com/topics/artificial-intelligence-business. [Accessed: Jun. 25, 2024].

[13] "5 ways Amazon is using AI to improve your holiday shopping and deliver your package faster," Amazon, Nov. 27, 2023 [Online]. Available: https://www.aboutamazon.com/news/operations/amazon-uses-ai-to-improve-shopping. [Accessed: Jun. 25, 2024].

[14] "What is artificial intelligence in medicine?," IBM, [Online]. Available: https://www.ibm.com/topics/artificial-intelligence-medicine. [Accessed: Jun. 25, 2024].

[15] S. Fujimori, "When AI Meets Biology," Pfizer, Jul. 2022, [Online]. Available: https://www.pfizer.com/news/behind-the-science/when-ai-meets-biology. [Accessed: Jun. 25, 2024].

[16] "Doing the math to pin down a rare heart disease," World Heart Federation, Sep. 21, 2020, [Online]. Available: https://world-heart-federation.org/news/doing-the-math-to-pin-down-a-rare-heart-disease/. [Accessed: Jun. 25, 2024].

[17] S. Chadha, "How digital helps a life sciences leader move at light speed," McKinsey & Company, May 31, 2022, [Online]. Available: https://www.mckinsey.com/industries/life-sciences/our-insights/how-digital-helps-a-life-sciences-leader-move-at-light-speed. [Accessed: Jun. 25, 2024].

[18] "Artificial intelligence (AI) cybersecurity," IBM, [Online]. Available: https://www.ibm.com/ai-cybersecurity. [Accessed: Jun. 25, 2024].

[19] "What is AI in cybersecurity?," Sophos, [Online]. Available: https://www.sophos.com/en-us/cybersecurity-explained/ai-in-cybersecurity. [Accessed: Jun. 25, 2024].

[20] "Stay Ahead with Cylance AI," BlackBerry, [Online]. Available: https://www.blackberry.com/us/en/products/cylance-endpoint-security/cylance-ai. [Accessed: Jun. 25, 2024].

[21] "Protect with AI," Microsoft, [Online]. Available: https://www.microsoft.com/en-us/security/business/solutions/generative-ai-cybersecurity#Products. [Accessed: Jun. 25, 2024].

[22] "Computer brains and autonomous driving," Mercedes-Benz, [Online]. Available: https://group.mercedes-benz.com/innovation/case/autonomous/artificial-intelligence.html. [Accessed: Jun. 25, 2024].

[23] D. Shapiro, "Mercedes-Benz Taking Vehicle Product Lifecycle Digital With NVIDIA AI and Omniverse," NVIDIA, Feb. 23, 2023, [Online]. Available: https://blogs.nvidia.com/blog/mercedes-benz-digitalization/. [Accessed: Jun. 25, 2024].

[24] "The Crop multiplier," CropX, [Online]. Available: https://cropx.com/. [Accessed: Jun. 25, 2024].

[25] "Watson Decision Platform for Agriculture." IBM, 2018, [Online]. Available: https://www.ibm.com/downloads/cas/0XZ0VRAW. [Accessed: Jun. 25, 2024].

[26] "Advancing education with AI," Google for Education, [Online]. Available: https://edu.google.com/why-google/ai-for-education/. [Accessed: Jun. 25, 2024].

[27] "Expanding Microsoft Copilot access in education," Microsoft, Dec. 13, 2023, [Online]. Available: https://www.microsoft.com/en-us/education/blog/2023/12/expanding-microsoft-copilot-access-in-education/. [Accessed: Jun. 25, 2024].

[28] "AI Dungeon," Latitude, [Online]. Available: https://play.aidungeon.com/. [Accessed: Jun. 25, 2024].

[29] "What are the different AI models in AI Dungeon?," Latitude, [Online]. Available: https://help.aidungeon.com/faq/what-are-the-different-ai-language-models. [Accessed: Jun. 25, 2024].

[30] "Nevermind," Affectiva, [Online]. Available: https://www.affectiva.com/success-story/flying-mollusk/. [Accessed: Jun. 25, 2024].

[31] Y. Maguid, "Assassin's Creed Origins – Building Living Worlds Through Artificial Intelligence," Ubisoft, May 2, 2018, [Online]. Available: https://news.ubisoft.com/en-us/article/1fBJdvSr4f7pMNH3EQ8Z2Y/assassins-creed-origins-building-living-worlds-through-artificial-intelligence. [Accessed: Jun. 25, 2024].

[32] R. Barth, "The convergence of AI and creativity: Introducing ghostwriter," Ubisoft, Mar. 21, 2023, [Online]. Available: https://news.ubisoft.com/en-us/article/7Cm07zbBGy4Xml6WgYi25d/the-convergence-of-ai-and-creativity-introducing-ghostwriter. [Accessed: Jun. 25, 2024].

[33] C. Lee, "AI plagiarism changers: How academic leaders can prepare institutions," Turnitin, Aug. 31, 2023, [Online]. Available: https://www.turnitin.com/blog/ai-plagiarism-changers-how-administrators-can-prepare-their-institutions. [Accessed: Jun. 25, 2024].

[34] P. Ceres and A. Hoover, "Kids are going back to school. So is ChatGPT," Wired, Aug. 23, 2023, [Online]. Available: https://www.wired.com/story/chatgpt-schools-plagiarism-lesson-plans/. [Accessed: Jun. 25, 2024].

[35] J. Nam, "56% of college students have used AI on assignments or exams," Best Colleges, Nov. 22, 2023, [Online]. Available: https://www.bestcolleges.com/research/most-college-students-have-used-ai-survey/. [Accessed: Jun. 25, 2024].

[36] M. Bohannon, "Lawyer Used ChatGPT In Court—And Cited Fake Cases. A Judge Is Considering Sanctions," Forbes, Jun. 8, 2023. [Online]. Available: https://www.forbes.com/sites/mollybohannon/2023/06/08/lawyer-used-chatgpt-in-court-and-cited-fake-cases-a-judge-is-considering-sanctions/. [Accessed: Jun. 25, 2024].

[37] R. Carpenter, "Cleverbot," Cleverbot, [Online]. Available: https://www.cleverbot.com/. [Accessed: Jun. 25, 2024].

[38] E. Jozuka, "Beyond dimensions: The man who married a hologram," CNN Health, Dec. 29, 2018, [Online]. Available: https://edition.cnn.com/2018/12/28/health/rise-of-digisexuals-intl/index.html. [Accessed: Jun. 25, 2024].

[39] "Gatebox," Gatebox.ai, [Online]. Available: https://www.gatebox.ai/. [Accessed: Jun. 25, 2024].

[40] Y. Obuno, "What happened to the Japanese man who 'married' virtual character Hatsune Miku?," The Mainichi, Jan. 11, 2022, [Online]. Available: https://mainichi.jp/english/articles/20220111/p2a/00m/0li/028000c. [Accessed: Jun. 25, 2024].

[41] S. Hinduja, "Generative AI as a Vector for Harassment and Harm," Cyberbullying Research Center, May 10, 2023, [Online]. Available: https://cyberbullying.org/generative-ai-as-a-vector-for-harassment-and-harm. [Accessed: Jun. 25, 2024].

[42] B. Cost and A. Court, "My girlfriend 'Claudia' was really an AI catfish — I feel cheated," New York Post, Apr. 12, 2023, [Online]. Available: https://nypost.com/2023/04/12/my-girlfriend-was-really-an-ai-catfish-i-feel-cheated/. [Accessed: Jun. 25, 2024].

[43] A. Seitz-Wald, "Finance worker pays out $25 million after video call with deepfake 'chief financial officer'," ABC News, Jun. 18, 2024. [Online]. Available: https://www.nbcnews.com/politics/2024-election/misleading-gop-videos-biden-viral-fact-checks-rcna133316. [Accessed: Jun. 25, 2024].

[44] K. Magramo, "British engineering giant Arup revealed as $25 million deepfake scam victim," CNN Business, May 17, 2024, [Online]. Available: https://edition.cnn.com/2024/05/16/tech/arup-deepfake-scam-loss-hong-kong-intl-hnk/index.html. [Accessed: Jun. 25, 2024].

[45] Trend Micro, "Unusual CEO fraud via Deepfake Audio Steals US$243,000 from UK Company," Trend Micro, Sep. 5, 2019, [Online]. Available: https://www.trendmicro.com/vinfo/us/security/news/cyber-attacks/unusual-ceo-fraud-via-deepfake-audio-steals-us-243-000-from-u-k-company. [Accessed: Jun. 25, 2024].

[46] L. Tung, "Forget email: Scammers use CEO voice 'deep-fakes' to con workers into wiring cash," ZDNet, Sep. 4, 2019, [Online]. Available: https://www.zdnet.com/article/forget-email-scammers-use-ceo-voice-deepfakes-to-con-workers-into-wiring-cash/. [Accessed: Jun. 25, 2024]

[47] K. Papachristou, "Revenge porn helpline annual report 2023." Revenge Porn Helpline, 2024, [Online]. Available: https://revengepornhelpline.org.uk/resources/helpline-research-and-reports/. [Accessed: Jun. 25, 2024].

[48] E. Clark, "Pedophiles using AI to generate child sexual abuse imagery," Forbes, Oct. 31, 2023, [Online]. Available: https://www.forbes.com/sites/elijahclark/2023/10/31/pedophiles-using-ai-to-generate-child-sexual-abuse-imagery/?sh=dd2cccf16564. [Accessed: Jun. 25, 2024].

[49] M. Murphy, "Predators exploit AI tools to generate images of child abuse," Bloomberg, May 23, 2023, [Online]. Available: https://www.bloomberg.com/news/articles/2023-05-23/predators-exploit-ai-tools-to-depict-abuse-prompting-warnings. [Accessed: Jun. 25, 2024].

[50] E. Casey, "People using AI to create child porn is a 'growing problem,' Wisconsin AG says," WPR, May 23, 2024. [Online]. Available: https://www.wpr.org/news/people-using-ai-to-create-child-porn-is-a-growing-problem-wisconsin-ag-says#:~:text=According%20to%20the%20U.S.%20Department%20of%20Justice%2C%20a%20child%20psychiatrist,into%20child%20sexual%20abuse%20material. [Accessed: Jun. 25, 2024].

[51] 'Fluffy89502,' "File:Map of the legality of child pornography by country.svg," Wikipedia, Apr. 24, 2020, [Online]. Available: https://en.wikipedia.org/wiki/File:Legality_of_child_pornography_by_country.svg. [Accessed: Jun. 25, 2024].

[52] IC3, "Child sexual abuse material created by generative AI and similar online tools is illegal," FBI, Mar. 29, 2024, [Online]. Available: https://www.ic3.gov/Media/Y2024/PSA240329. [Accessed: Jun. 25, 2024].

[53] B. Brittain, "Artists take new shot at Stability, Midjourney in updated copyright lawsuit," Reuters, Nov. 30, 2023, [Online]. Available: https://www.reuters.com/legal/litigation/artists-take-new-shot-stability-midjourney-updated-copyright-lawsuit-2023-11-30/. [Accessed: Jun. 25, 2024].

[54] E. Glover, "AI-Generated Content and Copyright Law: What We Know," Built in, Feb. 28, 2024. [Online]. Available: https://builtin.com/artificial-intelligence/ai-copyright#:~:text=For%20a%20product%20to%20be,work%20of%20a%20human%20creator. [Accessed: Jun. 25, 2024].

[55] Z. Sharf, "Marvel Used AI to Create 'Secret Invasion' Opening Credits, EP Says It Fits the 'Shape-Shifting' Plot," Variety, Jun. 21, 2023, [Online]. Available: https://variety.com/2023/tv/news/secret-invasion-artificial-intelligence-credits-marvel-1235650643/. [Accessed: Jun. 25, 2024].

[56] J. Weatherbed, "Disney's Loki faces backlash over alleged use of generative AI," The Verge, Oct. 9, 2023, [Online]. Available: https://www.theverge.com/2023/10/9/23909529/disney-marvel-loki-generative-ai-poster-backlash-season-2. [Accessed: Jun. 25, 2024].

[57] P. Bilderbeck, "People stunned as they spot entire row of AI extras in background of Disney show," Unilad, Oct. 13, 2023, [Online]. Available: https://www.unilad.com/film-and-tv/news/disney-prom-pact-ai-actors-851337-20231013. [Accessed: Jun. 25, 2024].

[58] Faciaai, "How did Paul Walker appear in Fast 7 after his death?" Medium, Nov. 27, 2023. [Online]. Available: https://faciaai.medium.com/how-did-paul-walker-appear-in-fast-7-after-his-death-fda6acfea096. [Accessed: Jun. 25, 2024].

[59] Guardian Staff, "Tom Hanks says AI version of him used in dental plan ad without his consent," The Guardian, Oct. 2, 2023, [Online]. Available: https://www.theguardian.com/film/2023/oct/02/tom-hanks-dental-ad-ai-version-fake?ref=upstract.com. [Accessed: Jun. 25, 2024].

[60] K. Tenbarge, "OpenAI pauses voice that many compared to Scarlett Johansson's," NBC News, May 20, 2024, [Online]. Available: https://www.nbcnews.com/tech/tech-news/openai-chatgpt-her-demo-video-scarlett-johannson-rcna153059. [Accessed: Jun. 25, 2024].

[61] D. Chmielewski and A. Tong, "Scarlett Johansson says OpenAI chatbot voice 'eerily similar' to hers," Reuters, May 21, 2024, [Online]. Available: https://www.reuters.com/technology/scarlett-johansson-says-openai-chatbot-voice-eerily-similar-hers-2024-05-21/. [Accessed: Jun. 25, 2024].

[62] SAG-AFTRA, "About - Mission Statement," SAG-AFTRA, [Online]. Available: https://www.sagaftra.org/about/mission-statement. [Accessed: Jun. 25, 2024].

[63] 'Robot Named Roy,' "AI Will Smith eating spaghetti pasta (AI footage and audio)," YouTube, Apr. 2023. [Online Video]. Available: https://www.youtube.com/watch?v=XQr4Xklqzw8. [Accessed: Jun. 25, 2024].

[64] R. Toews, "5 AI predictions for the year 2030," Forbes, Mar. 10, 2024, [Online]. Available: https://www.forbes.com/sites/robtoews/2024/03/10/10-ai-predictions-for-the-year-2030/. [Accessed: Jun. 25, 2024].

[65] K. Sachdeva, "Top 6 predictions for AI advancements and trends in 2024," IBM, Jan. 9, 2024, [Online]. Available: https://www.ibm.com/blog/top-6-predictions-for-ai-advancements-and-trends-in-2024/. [Accessed: Jun. 25, 2024].

[66] U. Wahid, I. Simpson, and N. Sargunaraj, "5 predictions: How AI, data privacy and cyber security could transform legal practices," KPMG, [Online]. Available: https://kpmg.com/xx/en/home/insights/2024/04/legal-predictions-on-data-privacy-cyber-security.html. [Accessed: Jun. 25, 2024].

Epilogue

And as such, dear reader, we must close the book on our adventure into the perils of the contemporary internet. Hopefully, you've laughed, maybe were a little confused, possibly even scared, and most likely changed your passwords half a dozen times. We started this book outlining the three predictions that were made when the world first went online. While the internet obviously did not disappear, we think it is fair to say that the utopian-flavored premonition didn't pan out too well. Despite the wonderful benefits of the internet, it has not been all clear and pleasant sailing. Humanity has not been magically transformed into a higher state of being, collectively singing kumbaya. Rather, throughout this book we have tried to reflect the idea that in many ways, the internet is really just a new context for fundamental human tendencies, dodgy behaviors, and old weaknesses to play out. And unfortunately, many of these are actually amplified by the ways in which common features of the internet (like social media) are designed.

We have seen, for example, how the internet has become a very effective conduit for increasingly creative con artists to ply what is a very old trade. One innocent click, and suddenly you're sending your life savings to "Danika" in eastern Ukraine to help stop her "cat hospital" from going into bankruptcy. If it's any consolation, you are not alone, and as you have seen, even those who think they are immune can easily get caught up too. In the future, take that extra second or two to check that URL you're about to click on. Better yet, the next time you see an outlandish claim, or even one that looks a bit more plausible, take a deep breath and maybe, just maybe, fact-check it before sharing it with your 300 closest friends. Might as well throw it through an AI-checker as well, as it may have just been generator by

DOI: 10.1201/9781032679389-12

everyone's favorite new author, ChatGPT. Do it like your life depended on it. Ohhh, and your bank balance will thank you later.

Once again, don't go around sharing your pet's name and your favorite color online, as while it may seem a harmless little gesture, your bank account may get cleaned out on the flip side. And then your credit score will resemble a limbo contest (lower, lower, lower). The moral of the story is simple: guard your personal info, trust no one, and triple-check everything online. And don't be afraid of a little bit more digital paranoia. It's definitely warranted given that your internet devices are almost certainly more attentive than your partner. Alexa and Siri are omnipresent and are like those friends who listen in on everything, but these digital friends will store it nicely away for their, or someone else's, future use. And be wary of the webcams in your own home too... we don't want another Thomas from Coventry.

David Bowie famously spoke on the internet's potential to change society, for both "good and bad," during an interview with the BBC in 1999. He seems to have gotten it right. Social networking platforms promised at the beginning to make our connections with others easier. They certainly have, but for many, it just ramped up the old human trait of social comparison. The internet turned into a place where everyone's life is perfect, except your own. Just remember, the grass is always greener on the internet, but that's usually because it's been photoshopped to perfection, especially on dating sites. As for kids, imagine for a moment that the future of society is like a game of Jenga, and they are the blocks at the top. We all know that things like poverty, bad nutrition, and abuse are like pulling out key pieces at the bottom, making the whole structure wobble. Yet we have been a lot slower to recognize that giving kids access to the internet in the way they currently have is like handing them a sledgehammer to whack away the rest of the remaining structure while saying, "have fun"! It's no surprise that the internet has become a hotspot for increasing numbers of kids with body image issues, anxiety, and dysfunctional sleep.

So, as we turn the final page and bid farewell to this exploration into the internet's labyrinthine landscape, remember that while the digital realm has its marvels, it also harbors shadows. We've walked you through the highs and lows, laughed at the absurdities, and shivered at the dangers lurking behind innocent clicks. Don't let the failed predictions of a utopian online world get you down, this doesn't diminish the internet's amazing and transformative power. It's just important to remember that it remains a mirror, reflecting both the brilliance and flaws of humanity. As you continue to navigate this digital frontier, just be vigilant. Protect your personal information, question the authenticity of whatever you encounter, and use that brain of yours. Foster a healthy skepticism. After all, the internet is still only a tool, powerful yet impartial (albeit not wholly... you can't blame the 1s and 0s for somebody's ulterior motives), and ultimately its impact lies in how humanity chooses to wield it. As always, with great power, comes great responsibility. Stay informed, stay cautious, and above all, stay curious. The adventure may end here, but your journey continues, shaped by each decision you make in this vast, ever-evolving world of the internet.

Acknowledgments

After nearly a year in the works, your three humble authors would like to pay tribute to those that got them to this point. Without these people's patience, support, and uncanny ability to distract us with Instagram reels at just the right moments, this book would have been finished much sooner—but far less enjoyably. Their belief in our writing kept us going, even when we were all unanimously convinced that the only people who would read this book would be... well us. Thank you all for putting up with the late-night typings, our endless mutterings and cursing, and the desperate searches for synonyms.

Marton:
I would like to thank my beautiful wife, Christy, and amazing daughter, Lana, for putting up with me... not just during the writing of this book, but just in general. I know that I'm usually not the most easy-going of folks (or am I?), but compound that with a huge looming deadline, 150 pages of manuscript writing, and nearly 300 pages of proofreading and editing, and I'm sure I was not exactly pleasant to be around. So, thank you both for being my rock, and my hard place.

Ian:
I dedicate this book to my best friend and life-limpet, Dr. Brettjet Lyn Cody. Thank you so much for the last year, the Helen Keller moments, the DT shuffles and shakes, the sofa surfing, and the constant, innumerable laughs. Without you, none of this would have been possible. You brought me back to life, so like the song goes, to distant lands, take both my hands.

Heba:

I dedicate this work to my family and friends that put up with my shit while working on this book. Bless their hearts. <3

I'd like to throw in the past, present, and potential victims of the issues mentioned in this book. Bless their hearts, but in a different way. c:

Index